JOURNAL OF CHROMATOGRAPHY LIBRARY — volume 47

trace metal analysis and speciation

JOURNAL OF CHROMATOGRAPHY LIBRARY — volume 47

trace metal analysis and speciation

edited by
I.S. Krull

Department of Chemistry, The Barnett Institute,
Northeastern University, Boston, MA 02115, U.S.A.

ELSEVIER
Amsterdam — Oxford — New York — Tokyo   1991

ELSEVIER SCIENCE PUBLISHERS B.V. **CHEMISTRY**
Sara Burgerhartstraat 25
P.O. Box 211, 1000 AE Amsterdam, The Netherlands

Distributors for the United States and Canada:

ELSEVIER SCIENCE PUBLISHING COMPANY INC.
655 Avenue of the Americas
New York, NY 10010, U.S.A.

ISBN 0-444-88209-X

PREFACE

This book arose out of a mini-symposium that was held at the Eastern Analytical Symposium, Fall, 1987, New York City, which I was asked to organize. As the lectures proceeded, it became clear that each speaker was an expert in his area of expertise, and that each had taken considerable time and effort to put together an outstanding talk. The attendance was above average and the talks were well received. Thus, half-way through, it occurred to the organizer that the collection of these lectures should not be lost to time. In discussing this with each lecturer individually, they were all receptive, enthusiastic, and excited about the idea of having their lectures, up-dated as time dictated, appear in book form for future interested readers. As more time developed, it became clear that several publishers were also quite interested in having this type of a book appear, and they all felt that there would be sufficient interest to warrant eventual publication. Few edited books have appeared dealing with trace metal analysis and speciation, which was the theme for the mini-symposium. Though there are any number of texts, edited and authored, which deal with trace metal analysis, few have emphasized trace metal speciation. That by van Loon appears to have come the closest, but even here it did not contain individual reviews of the most important (perhaps) areas of instrumental approaches for trace metal speciation. The Dahlem Conference of 1984 also came close to hitting the mark, for here there were many contributors, covering almost all the major areas of trace metal analysis and speciation. Though single chapters dedicated to this overall topic have appeared, mainly by a single author or a group of authors, a compilation of review chapters in a single text dedicated to trace metal speciation seems to have been overlooked and lacking. It was my intent to redress this oversight.

The current book, clearly edited, consists of nine separate, at times interconnected, chapters, each written by different authors. There are no two chapters by the same author(s), though I have encouraged individual authors to coordinate their reviews in order to avoid duplication of efforts. It is my impression that this has been successful. It has been my pleasure to have selected those individuals who eventually came through with their chapters, despite their very busy schedules and time pressures. I am indebted and grateful to all of them, and for their bearing up under my constant and insistent badgering for the final drafts, which eventually arrived. I am also grateful to the publishers who have been most helpful and cooperative, indeed overly patient, throughout negotiations, discussions, waiting periods, and the final publication periods. I am also grateful to several graduate students, especially Jeff Mazzeo and Andre Bourque at Northeastern University, who have been helpful in compiling the subject index for the final book.

Trace metal speciation refers to our ability to define which forms of a given metal or organometal appear in a particular sample, and at what precise quantitative levels such chemicals occur. Various methods have evolved, gas chromatography (GC)–mass spectrometry (MS), GC–plasma emission spectrometry, HPLC-UV/FL/EC detection, HPLC–mass spectrometry, HPLC–inductively coupled plasma (ICP) emission, HPLC-ICP-MS, GC/HPLC–direct current plasma (DCP) emission, and others. All of these have shown extreme promise for specific metal species in specific sample matrices, with varying degrees of selectivity, specificity, and sensitivity (as well as detection limits). They have all provided analysts with a choice of methods for trace metal species determination, and there is no clearcut formula for deciding which specific hyphenated or direct methods of analysis are best for a particular metal species in a particular sample matrix. Perhaps that is what is lacking in the overall scheme of trace metal speciation. Advice as to which method to choose for a particular metal species in a particular sample matrix would provide significant assistance to those really interested in applying the many methods available.

Chapter 1 by P.C. Uden provides an overview of the entire area of gas chromatography interfaced with various forms of element selective detection, especially plasma emission methods (MIP, DCP, ICP, etc.). He has provided us with a thorough review of the advantages, disadvantages, interfacing, optimization, and applications of GC–element selective detection for volatile metal/nonmetal species. Chapter 2 by K.J. Irgolic is a thorough review on the interfacing of various forms of HPLC with ICP, opportunities possible via HPLC-ICP, problems in detection limits, possibilities for metal speciation, and specific applications from the author's own experience and background. Chapter 3 by D.T. Heitkemper and J.A. Caruso reviews recent advances in plasma mass spectrometry, interfacing of GC and HPLC with ICP-MS techniques, and specific applications for trace metal speciation via GC/HPLC-ICP-MS. It is the state-of-the-art, and when combined with Chapter 5, they both provide a very complete and thorough picture of where ICP-MS and/or other forms of plasma emission-MS stand today. Chapter 5 by R.S. Houk and

S.-J. Jiang again deals with ICP-MS interfacing with HPLC and GC, and now draws on the author's own excellent background and experience in the initial development of ICP-MS as a stand-alone and hyphenated detector for chromatography. Houk has again described how the basic system works, problems in practice, ways to avoid and overcome problems of operation, ways to improve interfacing, and finally, specific applications at trace levels in real world samples.

Chapter 4 deals with improved instrumentation for plasma emission detectors, multichannel array detectors, the charge transfer detector, the charge injection device, CID echelle spectrometers, array detection of chromatographic eluents, and interfacing of chromatography with these ICP spectrometer designs for improved data acquisition, handling, manipulation, and interpretation. R.B. Bilhorn, R.S. Pomeroy and M.B. Denton have summarized almost a life-time's work in this one chapter, reviewing all that has been done in these areas of improved, intelligent spectrometers in atomic emission spectrometry, stand-alone or interfaced. Chapter 6 is an overview of newer HPLC approaches for trace metal cation separation and detection by B.D. Karcher and I.S. Krull. Coming from the Ph.D. Thesis of Karcher, it reviews the entire field of secondary chemical equilibria in dynamic ion-exchange and reversed phase for the separation of metal cation species. Separations are followed by metal complexation, post-column, together with UV and/or FL detection, with or without a liquid–liquid extraction technique. Special emphasis is placed here on the equilibria involved, kinetics of complex formation and dissociation, secondary chemical equilibria of metal species, formation of mixed metal complexes, formation of ion-paired species with mobile phase components, and final, overall chromatographic separation of the species present for a given metal cation in the form of their organic complexes. A separate computer program is described that has allowed the determination of elution orders for various metal cations under certain mobile phase conditions, approximate elution times, resolutions, separation (alpha) values, and related chromatographic figures of merit. Chapter 7 by A.M. Bond is an overview of how electrochemical detection in liquid chromatography can be used with pre-column, automated or manual formation of metal complexes for trace metal speciation of real world, environmental water samples on a continuous, day-to-day basis in Australia. Presumably, these methods will work up-over, in the Northern Hemisphere, as well as they do down-under. However, the orders of elution of metal complexes under reversed phase conditions may change in going from Australia to the United States, because of changes in the earth's magnetic field directions. Nevertheless, Bond has described a beautifully optimized and engineered instrument for performing routine, in-the-field, daily determinations of metal species levels in real world environmental samples. The overall results have been very impressive, and it is perhaps an excellent example of how academic research into HPLC-EC methods for trace metal determinations has been applied on a practical level for real world problem solving, with a high degree of automation and success. Chapter 8 discusses ion chromatographic approaches for trace metals, and written by some of the early pioneers in this area, D.T. Gjerde and H.C. Mehra, it is a very complete overview of how various forms of IC can today be used to perform true trace metal speciation. Many, if not most, of the more significant environmental metals have been covered, and the reference list is quite exhaustive and complete. Although IC does not usually offer very specific detection methods, it can and has done that, especially when interfaced with plasma emission detection. Finally, Chapter 9 by I.S. Krull and Wm. Childress deals with the use of direct current plasma emission spectroscopy/detection for trace metal speciation in both GC and HPLC modes. This chapter provides an overview of much/most of the literature that has used DCP in chromatography, and emphasizes work performed within Boston or Winchester Food & Drug Administration laboratories during the past ten years (1980s). This particular chapter may prove useful because it provides real world sample analysis, such as methylmercury in swordfish or tributyltin in shellfish, areas of current, intense interest and significance. Our ability to simply, quickly, and inexpensively interface both GC and HPLC with commercial DCP sources should encourage others to attempt such hyphenated techniques for future trace metal analysis and speciation in other real samples. Quantitation is clearly practical, accurate, precise, and very reproducible, even with actual samples from the environment.

I wish to indicate my sincere appreciation to the contributors of this book, for all of the hard work and toil that has gone into each and every chapter, and for tolerating the incessant urgings of the Editor to finish their voluntary tasks. By and large, this book has been a labor-of-love, and though it has taken more time than one would have liked to complete, it is hoped that the final product will be accepted as worthy of all the collective efforts. Reviewers, be kind!

May, 1991
IRA S. KRULL
Boston, Massachusetts, USA

LIST OF CONTRIBUTORS

R.B. BILHORN   *Eastman Kodak Company, Kodak Park Building #34, Rochester, NY 14650, U.S.A.*

A.M. BOND   *Department of Chemical and Analytical Sciences, Deakin University, Geelong, Victoria 3217, Australia*

J. CARUSO   *Department of Chemistry, University of Cincinnati, Cincinnati, OH 45221 U.S.A.*

Wm. CHILDRESS   *Analytical Chemistry Branch, U.S. Food and Drug Administration, Winchester Engineering and Analytical Chemistry, Winchester, MA 01890, U.S.A.*

M.B. DENTON   *Department of Chemistry, University of Arizona, Tucson, AZ 85721, U.S.A.*

D.T. GJERDE   *Sarasep, Inc. 1600 Wyatt Drive, Suite 10, Santa Clara, CA 95054, U.S.A.*

D.T. HEITKEMPER   *Elemental Analysis Research Center, U.S. Food and Drug Administration, Cincinnati, OH 45202 U.S.A.*

R.S. HOUK   *Ames Laboratory, U.S. Department of Energy, and Chemistry Department, Iowa State University, Ames, IO 50011, U.S.A.*

K. IRGOLIC   *Department of Chemistry, Texas A&M University, College Station, TX 77843, U.S.A.*

S.-J. JIANG   *Chemistry Department, National Sun Yat-Sen University, Kaohsiung Taiwan 80424, Republic of China*

B.D. KARCHER   *The Barnett Institute of Chemical Analysis and Materials Science, Northeastern University, Boston, MA 02115, U.S.A.*

I.S. KRULL   *Department of Chemistry and The Barnett Institute of Chemical Analysis and Materials Science, Northeastern University, Boston, MA 02115, U.S.A.*

H.C. MEHRA   *Chemical Waste Management, Inc., P.O. Box 4249, Modesto, CA 95352, U.S.A.*

R.S. POMEROY   *Department of Chemistry, University of Arizona, Tucson, AZ 85721, U.S.A.*

P.C. UDEN   *Department of Chemistry, University of Massachusetts at Amherst, Amherst, MA 01003, U.S.A.*

CONTENTS

I.S. Krull (Ed.), *Trace Metal Analysis and Speciation*
Journal of Chromatography Library Series, Vol. 47
© 1991 Elsevier Science Publishers B.V., Amsterdam

1

Chapter 1

Chromatography-Element Selective Detection: Interfacing of Gas Chromatography with Microwave-Induced Plasma Emission Detection (GC-MIP)

P.C. UDEN

Department of Chemistry, Lederle Graduate Research Tower, University of Massachusetts, Amherst, MA 01003, U.S.A.

1.1. INTRODUCTION

The advances in recent years in analytical atomic spectroscopy, particularly the renaissence of atomic emission methods utilizing plasma sources, have greatly expanded capabilities for trace metals analysis. While determination of the total elemental content of an analyte remains one of the most frequent objectives, increasing emphasis has been placed upon "speciation" of analytes on an elemental basis; that is, the full characterization of the chemical nature of the analyte molecule in which a particular target element is present. In this regard, sample separation methods, particularly those involving chromatography, are an inherent part of the analytical process.

Chromatography has as its basis the transformation of a complex multi-component sample into a time-resolved, separated analyte stream, usually observed in the analog differential signal mode. The "chromatographic sample" is thus distinctive in that analytes are changing in nature and with time. An essential feature of chromatographic instrumentation is a detector for qualitative and quantitative determination of the components resolved by the column; this should respond immediately and predictably to the presence of solute in the mobile phase. An important class of solute property detectors are those giving "Selective", or "Specific" information on the eluates. Spectral property detectors such as the mass spectrometer, the infrared spectrophotometer and the atomic emission spectrometer fall into this class. Such detectors may be "element selective", "structure or functionality selective" or "property selective".

It is worth noting the different perspectives on an "instrumental" system which interfaces the chromatograph with a complex "sample characterization device" such as the atomic emission spectrometer. To the chromatographer, the spectrometer is a sophisticated "chromatographic detector"; from the spectroscopists point of view, the chromatograph is a component-resolving sample introduction device. Both perspectives are accurate and emphasize that this mode of chemical analysis requires optimization of both the separation and the detection process, together with the "interface" linking them.

1.1.1. Element selective gas chromatographic detection

Element selective chromatographic detection aims to obtain qualitative, and quantitative analysis of eluates, frequently in interfering background matrices, based upon their elemental constitution. Simultaneous multielement detection may also permit the determination of the empirical formulae of eluates. Element selective gas chromatographic detectors in common use include the alkali flame ionization detector (AFID), often known as the nitrogen/phosphorus detector (NPD), selective for these elements; the flame photometric detector (FPD), selective for sulfur and phosphorus, and the Hall electrolytic conductivity detector, which is selective for halogen, nitrogen and sulfur. The wide application of these detectors shows the value of element selective detection, but they cannot be used for general eluate empirical formula determinations. No specific commercially available GC detectors for metals have been introduced, with the exception of the atomic plasma emission instruments discussed later. It is evident that general multielement chromatographic detection is a worthwhile technical objective to compliment the molecular and structural specific detection afforded in GC by interfaced mass spectroscopy and Fourier transform infra-red spectroscopy. Atomic emission spectroscopy is a natural choice for such interfaced detection in view of its capacity to monitor all elements.

1.1.2. GC-atomic emission detection (GC-AED) of metals

Atomic emission spectral detection of metallic elements in GC shows something of an irony; while detection limits and selectivities are usually better than for non-metals, because of greater emission intensities and absence of background interferences in the spectral region monitored, GC applications for metallic compounds in general are less common than for non-metals. However, many volatile binary metal and metalloid compounds, organometallic and metal chelate compounds can be quantitatively gas chromatographed [1], and some of the possible GC-

AED detection methods are most valuable in confirming elution and acquiring sensitive analytical data.

1.2. ATOMIC EMISSION SPECTROSCOPIC DETECTION (AESD) IN CHROMATOGRAPHY

Four modes of atomic spectroscopy have been interfaced for chromatographic detection, flame emission (FES), atomic absorption (AAS), atomic fluorescence (AFS), and atomic plasma emission (APES) [2–3]. In contrast to AAS, APES can accomplish simultaneous multielement determination, while giving a good dynamic measurement range and high sensitivities and selectivities over background elements in many cases.

The development of many alternative plasma sources has led to a resurgence of analytical atomic emission spectroscopy in recent years. The major plasma emission sources used for gas chromatographic detection have been the microwave-induced helium plasma, under atmospheric or reduced pressure (MIP), and the DC argon plasma (DCP). The inductively coupled argon plasma (ICP) has been used much less for GC than as an HPLC detector [4].

The principal advantages of interfaced chromatography–atomic plasma emission spectroscopy (C-APES) are;

(a) monitoring eluted species for their elemental composition with high elemental sensitivity,

(b) monitoring for molecular functionality by derivatization with reagents incorporating specific "tag" elements,

(c) toleration of incomplete chromatographic resolution, the specificity of plasma emission enabling incomplete separation from complex matrices to be overcome,

(d) simultaneous multi-element detection for empirical and molecular formula determination,

(e) compatibility with different chromatographic systems, mobile phases and conditions.

1.2.1. Goals of atomic plasma emission chromatographic detection

1.2.1.1. *Inter-element selectivity*

Detection of the target element without interference or signal contribution from other elements present in the plasma is most important.

Such selectivity depends on the emission spectrum of the element and of possible interferences, and on the characteristics of the spectroscopic measurement system. Some UV-visible spectral regions are less liable than others to interference from the helium or argon plasma background, or from line or band spectra of carbon, nitrogen, oxygen or molecular combinations. A useful measure of inter-element selectivity is the peak area response per mole of analyte element divided by the peak area response of the "background" element per mole of that element, measured at the emission wavelength. Selectivity against carbon is usually measured, but other elemental background matrices have their own selectivity criteria. Selectivities vary between plasmas and with instrumental conditions, so calibration is always required. Chromatographic behavior and plasma emission response for each element must be linear; otherwise unjustifiably high selectivities may be found.

1.2.1.2. Elemental sensitivity and limits of detection

Element sensitivity depends on the emission intensity at the measured wavelength. Each element has a number of possible wavelengths for determination and the best must be chosen also taking into account selectivity. Different emission lines have different sensitivities in different plasmas. Sensitivity, defined by the slope of the response curve, is less often used in C-AED than "detection limit", expressed as absolute values of element mass (in a resolved peak) or in mass flow rate units. Detection limits for different elements vary by two or three orders of magnitude; this affects inter-element selectivity if spectral overlap is present.

1.2.1.3. Dynamic measurement range

Linear dynamic response ranges for capillary GC-AED extend from the upper linear analyte-carrying column capacity, at ca.100 ng, down to the detection limit for the element (1–100 pg). Chemical, dopant gas and plasma-wall interaction effects modify the limits.

1.2.1.4. Simultaneous multi-element detection

Specific multi-element monitoring is possible in various ways. Sequential switching between elemental wavelengths of a monochromator is useful if speed is compatible with peak elution rates [5]. However, the most widely utilized multi-element detection has been by means of a direct reading polychromator which can display up to 12 monitoring wavelengths simultaneously [6, 7]. This has been used with either reduced pressure [7] or atmospheric pressure MIP detection [8, 9]. Photodiode-array detection has shown considerable versatility and sensitivity [10], and a commercial instrument has been introduced [11], incorporating an atmospheric pressure MIP, a water-cooled discharge tube to improve signal to background signals, and a movable diode array measuring from 170–780 nm. The array range is approximately 25 nm, which defines which combinations of elements can be measured at one time. Among the attractive features of multi-channel GC-AED detection is quantitative element ratioing to give empirical formulae of eluates. Such determinations do not attain the accuracy of classical milligram level microanalysis, and presuppose no response dependence on molecular structure. However, measurements are made directly on eluent peaks at sample levels up to six orders of magnitude below classical methods.

1.2.2. Classes of atomic plasma emission chromatographic detectors

An emission spectral excitation source transforms a sample from a solid, liquid or gas into an energetic plasma of electrons, atoms, ions and radicals which can be electronically excited. When the excited states deactivate, they generate light quanta producing an elemental emission spectrum. The major classes of analytical plasmas are the Inductively Coupled Argon Plasma (ICP); the DC Argon Plasma (DCP), the Microwave-Induced Plasma (MIP), the Alternating Current Plasma (ACP) and the Capacitively Coupled Plasma (CCP). Of these that most frequently applied as a GC detector has been the MIP; this will be the system focussed upon in this discussion.

1.2.3. The microwave-induced electrical discharge plasma (MIP) detector

An argon or helium plasma is sustained in a microwave "cavity" which serves to focus or couple power from a microwave source, usually operated at 2.45 GHz, into a discharge cell which is a capillary tube, made of quartz, boron nitride, alumina etc.. Microwave plasmas of different cavity designs may be operated at atmospheric or under reduced pressures [12, 13]. The 50–100 watt power levels for analytical microwave plasmas are much lower than for the DCP or the ICP, giving

6

easier operation. Power densities are similar, however, since the MIP is much smaller. Although plasma temperatures are somewhat lower than for some other plasmas, high electron temperatures exist, particularly in the helium plasmas, giving intense spectral emission for many elements, including non-metals, which respond poorly in the argon ICP or DCP. Helium MIP systems have proved less useful for liquid chromatographic introduction since there is insufficient plasma enthalpy to desolvate and vaporize aerosols effectively except at energies approaching those of the ICP or DCP. A comparison of microwave cavities was made by Risby and Talmi in their general review of GC-MIP [14].

The most commonly used cavity for reduced pressure helium or argon plasmas is the 3/4 wave cavity described by Fehsenfeld et al. [15], but the cavities most widely adopted for GC-MIP are based upon the TM_{010} cylindrical resonance cavity developed by Beenakker [16] (Fig. 1.1). This cavity can sustain helium or argon atmospheric pressure discharges at power levels from 30 watts up. Emitted light is viewed axially and not transversely through the cavity walls whose transmission properties change with time. The discharge tube can be made from opaque materials such as alumina or boron nitride to provide improved performance for some elements. The absence of a vacuum system is an operational advantage.

Another type of microwave plasma cavity which has been used successfully in GC-MIP is the "Surfatron" which operates by surface microwave propagation along a plasma column [17]. The plasma may be viewed axially or transversely since it extends outside the plasma structure; it can sustain a dicharge over a wide pressure range.

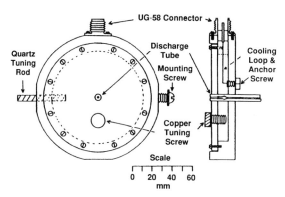

Fig. 1.1. The Beenakker microwave plasma cavity. (Reproduced by permission, from *J. Chromatog.*, 1982, 239, 181; copyright 1982, Elsevier Science Publishers.)

Since eluent from GC columns is at atmospheric pressure, simpler interfacing configurations are possible with atmospheric pressure plasmas than for reduced pressure plasmas. Interfacing of reduced pressure MIPs with GC involves evacuating a silica sample chamber inside the MIP cavity to a pressure of ca. one torr [18]. With packed columns, little loss in peak efficiency occurs, but the volume of the cavity produces inevitable broadening and tailing of capillary peaks. The atmospheric pressure cavities such as the TM_{010} are simple to interface with flexible fused silica capillary columns since these can be threaded towithin a few millimeters of the plasma, leading to minimal "dead volume". Helium make-up gas or other reactant gases can be introduced within the heated transfer line to optimize plasma performance and further minimize peak broadening. A typical GC-MIP interface is illustrated [19] (Fig. 1.2).

GC-MIP performance has been improved with a threaded tangential flow torch (TFT) [20, 21], to give a self-centering plasma which can give enhanced emission and stability. The plasma loses relatively little energy to the walls, thus atom formation and excitation appear to be enhanced by comparison with the straight capillary torch. Further, interference due to wall reactions or elements present in the walls can be reduced. A disadvantage, however, is the high volume (liters per minute) of helium flow gas required.

Interfacing TM_{010} cavities with packed GC columns is more difficult since the plasma is extinguished by the vapor burst from the solvent and may also be disrupted by large sample peaks. Packed columns are of value, however, particularly for trace determinations when resolution

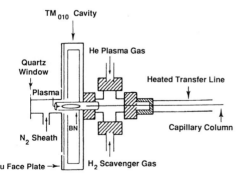

Fig. 1.2. GC-MIP Interface. (Reproduced by permission, from C. Bradley and J.W. Carnahan, *Anal. Chem.*, 1988, 60, 858; copyright 1988, American Chemical Society.)

References pp. 19–20

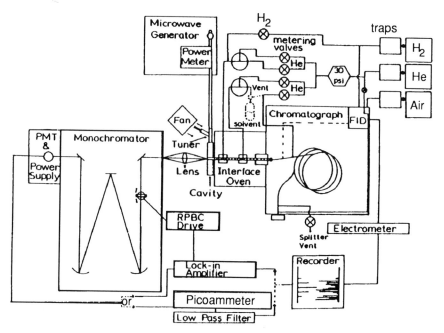

Fig. 1.3. An instrument system for GC-MIP, including a fluidic logic interface. (Reproduced by permission, from S.A. Estes, P.C. Uden and R.M. Barnes, *Anal. Chem.*, 1981, 53, 1336; copyright 1981, American Chemical Society.)

is adequate. Various techniques have been used to address this problem. The plasma may be re-ignited after the solvent peak has passed, or a valve or other venting system can be used to divert larger peaks away from the plasma. One system employed a "fluidic logic" flow system within an interface oven between the column and the MIP cavity [22] (Fig. 1.3). Similar valve systems have also been used to interface an additional parallel or series detector for independent analysis.

1.3. MIP DETECTION FOR HPLC AND SFC

The microwave-induced plasmas (MIP) have been much more effective in GC than in HPLC interfacing, the latter having been developed most with the high powered DCP and ICP argon plasmas which are able to tolerate mobile phase solvents at the flow rates used in liquid chromatographic procedures. The low powered helium MIP cannot be effectively interfaced to conventional HPLC columns, since the discharge will be quenched by continuous liquid flow streams Some approaches to this problem have been explored. Direct introduction of liquid into an MIP has been achieved by flowing the HPLC effluent over a heated wire and vaporizing it into the discharge by a cross-stream of helium. This procedure showed some potential for reversed phase HPLC [23]. A mixed gas oxygen–argon MIP maintained in a modified discharge tube consisting of two concentric quartz tubes was applied for HPLC of mercury compounds using methanol/water mobile phases; detection limits for organically-bound Hg were in the ng range [24]. A high-power (kW) discharge, operating in the radio frequency or microwave range, accommodated continuous solvent-flows [25]. The direct injection nebulizer (DIN) provides another approach. This device [26] transfers mobile phase flows of up to 0.5 ml min^{-1} into the plasma with efficiencies approaching 100%. A moving-wheel sample transport–desolvation system has been described in which aqueous solvent is evaporated by hot nitrogen, leaving dry analyte to be transported into a small volume 100W plasma, where it was volatilized, atomized and excited [27]. Other solvent-removal devices such as the "Thermospray" and "Particle-beam" approaches now used in HPLC-MS could be adapted since the interfacing problems for lower-powered MIPs are parallel to those in HPLC-MS and HPLC-FTIR. Another possibility is cryo-focusing as used in the latter technique. Alternative plasma cavities may become available to sustain a helium MIP under conventional HPLC flow conditions. Capillary HPLC columns with mobile phase flow rates of a few milliliter per minute provide an interesting possibility for helium MIP interfacing, but sample capacity may limit application for trace determinations.

Although analytical SFC was demonstrated in the early 1960s, it has only been in recent years that the availability of adequate high resolution packed and capillary SFC columns and instrumentation has led to renewed interest in the technique. Plasma emission is a natural development because of its use in GC and HPLC. A surfatron MIP sustained in helium has been employed for SFC detection, giving sulfur-specific detection at 921.3 nm with a 25 pg s^{-1} limit for thiophene [28]. An argon high efficiency MIP has been interfaced with packed column SFC and the separation and detection of ferrocene and derivatives achieved with iron specific detection. Methanol modifier concentrations to 5% were tolerated in the carbon dioxide mobile phase [29].

1.4. GC-MIP DETECTION OF METALS

1.4.1. Reduced pressure plasmas

As noted earlier, the reduced pressure 3/4 wave cavity [15] has been the most widely used for helium and argon plasmas for GC-MIP, although most applications have been for non-metallic elements. However, some notable applications for metal specific detection have been reported. Dagnell et al. [30] first used this cavity for the detection of aluminum, chromium, copper, gallium, iron, scandium and vanadium, gas chromatographed on packed glass columns as their acetylacetonates or trifluoroacetylacetonates. Limits of detection ranged between 2×10^{-11} g and 2×10^{-12} g, and selectivities against carbon were between 1000 and 4000. Chromium in blood plasma was determined as its trifluoracetylacetonate, with excellent quantitation and precision [31].

Trivalent arsenic and antimony were determined in environmental samples by derivatization and reaction to form stable triphenylarsine and triphenylstibine; these were extracted, gas chromatographed and detected at 228.8 nm and 259.8 nm respectively, with detection limits of

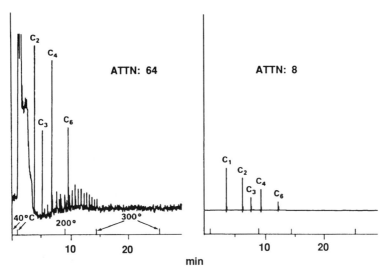

Fig. 1.4. Comparison of atmospheric pressure (left) and low pressure (right) GC-MIP for dialkylmercury compounds in a shale oil matrix. Mercury-specific detection at 253.6 nm. C_1, C_2, ... refer to dimethylmercury, diethylmercury, etc. (Reproduced by permission from K.B. Olsen, D.S. Sklarew and J.C. Evans, *Spectrochim. Acta.*, 1985, 40B, 357; copyright 1985, Pergamon Press.)

20 and 50 pg [32]. A parallel study [33] determined alkylarsenic acids in commercial pesticides and environmental samples by borohydride reduction. Reamer et al. determined tetraalkyl lead compounds in the atmosphere after collection in a cold trap [34]. Hydrogen was doped into the plasma to minimize lead build-up on the quartz plasma tube walls. A typical detection limit for tetraethyl lead was 6 pg, and selectivity over carbon was ca. 80,000.

Sklarew et al. [35] have noted that while the low pressure Evenson source provides optimal excitation energy for non-metals, its sensitivity is inherently less than that of the Beenakker cavity because of its transverse viewing geometry. The atmospheric pressure Beenakker cavity provides a more efficient way to couple microwave energy into the plasma, and with its axial geometry is inherently more sensitive; however its high pressure causes tube erosion, at least if the discharge tube is not cooled, thus causing reliability problems. Olsen et al. [36] compared reduced pressure and atmospheric pressure MIP systems for Hg, Se and As detection in organometalloid compounds in shale oil matrices, and found the latter to be superior both in detection limits and in selectivities. Figure 1.4 shows comparison chromatograms for low pressure and atmospheric pressure helium plasmas, of dialkylmercury compounds in a shale oil matrix.

1.4.2. Atmospheric pressure plasmas

Increased transfer efficiency of microwave power to the discharge using such cavity structures as the Beenakker TM_{010}, allows plasmas to be sustained at atmospheric pressure at the low power levels possible with reduced pressure cavities. A further advantage is the ability to view light emitted axially from the plasma, rather than transversely through the discharge tube wall with the resultant degradation in response with extended use. Polyimide coated flexible fused silica capillary (FSOT) columns can be interfaced to within a few mm of the plasma, and such direct interfaces have been widely used in capillary GC-MIP, although there are some advantages in incorporating a gas switching device. Such a system, including a deactivated valveless fluidic logic device allowing solvent venting and addition of dopant gas, was used successfully to transfer chemically active and thermally sensitive trialkyl lead chlorides at the sub-nanogram level [22]. Figure 1.5 shows chromatograms of triethyl lead chloride extracted into benzene from tap water, followed by concentration by vacuum reduction, with a detection limit of ca. 20 ppb. The practical limits of this procedure lie in the assurance

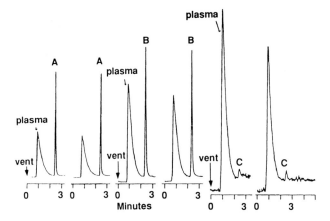

Fig. 1.5. Extraction of Et₃PbCl from spiked tap water. Pb I monitored at 405.8 nm. Replicate samples (2) for *A*, 5.8 ppm, *B*, 400 ppb and *C*, 34 ppb in tap water. (Reproduced by permission from S.A. Estes, P.C. Uden and R.M. Barnes, *Anal. Chem.*, 1981, 53, 1336; copyright 1981, American Chemical Society.)

of quantitative transfer of these reactive analytes to the plasma, and exhaustive deactivation of the quartz transfer system was needed.

As shown in Table 1.1, GC-MIP data has been obtained for many transition and main group metals. Many of the analytical figures of merit reported in this compilation were obtained with the fluidic logic interface noted above and are abstracted from the tabulation in a comprehensive evaluation of MIP emission detection characteristics, which also lists the model compounds examined by fused silica capillary GC [37]. It should be noted that some of the quoted detection limits, notably those for arsenic, niobium, tungsten and indium are affected by non-ideal chromatographic behavior of the analytes towards the column, as well as by thermal instabilities and reaction with carrier gas impurities. These are all problems inherent in metal compound gas chromatography, rather than being due to any shortcomings of the plasma atomic emission process and measurement. Some of the metals, notably lead and mercury, have been the subject of a number of studies and each is determinable by GC-MIP with TM_{010} cavities to sub-pg s^{-1} detection limits.

In a comparison investigation to that noted above, lead and carbon specific detection were compared for trialkyllead chlorides extracted from an industrial plant effluent and derivatized with butyl Grignard reagent to form their analogous trialkylbutyllead compounds [38]. Chromatograms are shown in Fig. 1.6. The degree of chromatographic inter-

TABLE 1.1

SELECTED DETECTION LIMITS AND SELECTIVITIES FOR ATMOSPHERIC PRESSURE HELIUM MICROWAVE PLASMA GC DETECTION[a]

Element	Wavelength (nm)	Detection limit pg s^{-1}	(pg)	Selectivity vs. C	Dynamic range
Carbon (b)	247.9	2.7	(12)	1	1,000
Boron (b)	249.8	3.6	(27)	9,300	500
Sulfur (a)	180.7	2		8,000	10,000
Phosphorus (a)	177.5	1		5,000	1,000
Silicon (b)	251.6	9.3	(18)	1,600	500
Tin (b)	284.0	1.6	(6.1)	36,000	1,000
Selenium (b)	204.0	5.3	(56)	11,000	1,000
Arsenic (b)	228.8	6.5	(155)	47,000	500
Vanadium (b)	268.8	10	(26)	57,000	100
Chromium (b)	267.7	7.5	(19)	108,000	1,000
Manganese (b)	257.6	1.6	(7.7)	110,000	1,000
Iron (b)	259.9	0.3	(0.9)	280,000	1,000
Cobalt (b)	240.7	6.2	(18)	190,000	1,000
Nickel (b)	231.6	2.6	(5.9)	6,500	1,000
Niobium (b)	288.3	69	(335)	32,000	100
Molybdenum (b)	281.6	5.5	(25)	24,000	500
Ruthenium (b)	240.3	7.8	(35)	134,000	1,000
Rhodium (d)	343.5		(211)	1,100	1,000
Tungsten (b)	255.5	51	(646)	5,500	500
Osmium (b)	225.6	6.3	(34)	50,000	1,000
Mercury (b)	253.7	0.6	(60)	77,000	1,000
Aluminum (b)	396.2	5.0	(19)	3,900	500
Gallium (c)	294.3		(200)	500	1,000
Germanium (b)	265.1	1.3	(3.9)	7,600	1,000
Indium (c)	325.6		(8,000)	600	1,000
Tin (b)	284.0	1.6	(6.1)	36,000	1,000
Lead (b)	283.3	0.17	(0.71)	25,000	1,000

[a] Compiled from (a) [11], (b) [37], (c) [44] and (d) [48].

ference from the high level of carbon-containing compounds would prevent any qualitative or quantitative determination of the trialkyl lead compounds by flame ionization detection, GC-ECD or GC-MS without extensive clean-up and loss of analyte.

In their study comparing reduced and atmospheric pressure MIP detection, Olsen et al. found for the latter system a one pg detection limit for mercury, with selectivity over carbon of 10,000; dilakyl mercurials, diarylmercurials, monoalkyl mercury chlorides, trialkyl arsines and triarylarsines were all chromatographed [36]. GC-MIP of volatile

14

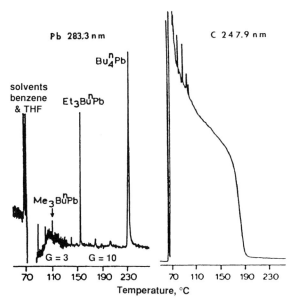

Fig. 1.6. Simultaneous atmospheric-pressure GC-MIP chromatograms showing lead (left) and carbon (right) detection of industrial plant water effluents derivatized as tributylalkyl-leads. (Reproduced by permission, from S.A. Estes, P.C. Uden and R.M. Barnes. *Anal. Chem.*, 1982, 54, 2402; copyright 1982, American Chemical Society.)

elemental hydrides of germanium, arsenic, selenium, tin and antimony gave sub-ng detection [39]. Simultaneous detection by a direct reading polychromator system enabled measurements of these hydrides, generated from solution by borohydride reaction, to be made, and there is considerable potential for such determinations in environmental matrices.

A commercial GC instrument has been developed [11], incorporating an atmospheric pressure helium MIP which uses a water-cooled discharge tube to maximize signal to background signals, and a moveable photodiode array detector measuring from 170 to 780 nm. Simultaneous detection of up to four elements and display of element-specific chromatograms is available. The array range is approximately 25 nm, which dictates which combinations of elements can be measured in a single experiment. The system also displays an elemental emission spectrum of the type shown in Fig. 1.7 for tin, obtained here on 10 pg of the element.

Metal chelates of sufficient volatility and thermal stability for gas chromatography have been extensively studied over the past 20 years,

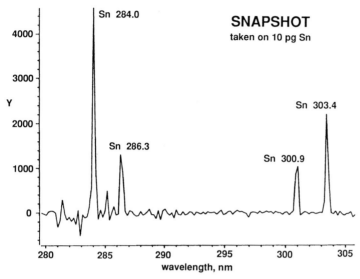

Fig. 1.7. Instantaneous spectrum of the GC-AED photodiode array wavelength range for tin detection, showing primary and secondary emissions. (Reproduced by permission from R.L. Firor, *American Lab.*, 1989, 21(5), 1989; copyright, 1989, International Scientific Communications, Inc.)

with most emphasis being placed on complexing ligands of 2,4-pentane-dione (acetylacetone) and its fluorinated derivatives; this area has been comprehensively reviewed [1, 40–43]. The GC-AED detection of chelates has a number of valuable features. In many instances, notably for complexes of fluorinated diketones with trivalent hexa-coordinate main and transition group metals, such as aluminum and chromium, and of tetradentate fluorinated ketoamine complexes of divalent transition metals such as copper and nickel, excellent GC behavior with high resolution and quantitative elution has been achieved. This has enabled trace metal determination by derivatization procedures and also has given valuable model compounds for assessment of element specific MIP conditions and behavior for calibration purposes. A recent study utilizing atmospheric pressure GC-MIP detection involved ligand redistribution and reaction kinetics of gallium, indium and aluminum chelates [44]. Figure 1.8 shows a capillary GC study of redistribution of trifluoroacetylacetone (TFA) (1,1,1-trifluoropentane-2,4-dione) and TIB (1,1,1-trifluoro-6-methylheptane-2,4-dione) ligands on gallium and aluminum atom centers. Al(TFA)$_3$ and Ga(TIB)$_3$ were reacted in chloroform solution for 24 h at 25°C and the products examined with carbon, aluminum and gallium specific GC detection. Individual peaks were

References pp. 19–20

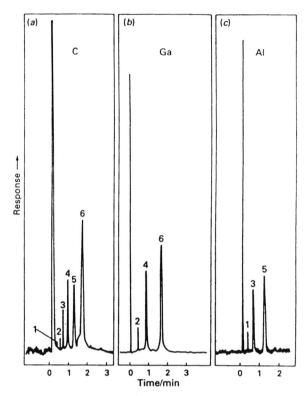

Fig. 1.8. GC-AED of redistribution of ligands on gallium and aluminum. Column 5 m × 0.2 mm i.d. FSOT SE 30 capillary at 108°C. (a) Carbon detection at 247.9 nm, (b) gallium detection at 294.3 nm, (c) aluminum detection at 396.1 nm. (Reproduced by permission from *J. Analytical Atomic Spec.*, 1988, 3, 919; copyright 1988, Royal Society of Chemistry.)

identified as $Al(TFA)_2(TIB)$ — 1; $Ga(TFA)_2(TIB)$ — 2; $Al(TFA)(TIB)_2$ — 3; $Ga(TFA)(TIB)_2$ — 4; $Al(TIB)_3$ — 5 and $Ga(TIB)_3$ — 6. The most volatile trifluoroacetylacetonates were not observed. In addition to affording a way to obtain reaction kinetics and other physicochemical data for such reactions, the ability of GC-MIP to monitor such metals as Al, Ga and In offers good possibilities for analysis of these elements in their volatile organometallic compounds used in semi-conductor and electronic materials production.

Many Pi-bonded organometallics such as metallocene derivatives have proved well-behaved in capillary GC and have proved to be excellent model compounds for the evaluation of GC-MIP detection for their constituent elements. Estes et al. employed a TM_{010} cavity which gave

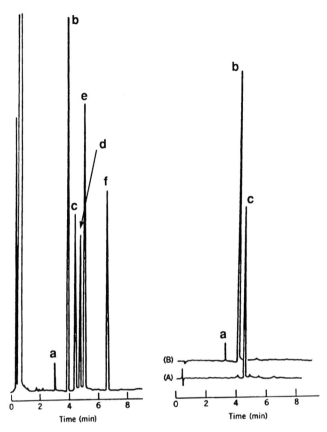

Fig. 1.9. Atmospheric pressure microwave GC-AED of organometallics. Left: carbon monitored at 247.9 nm; right: A, chromium monitored at 267.7 nm and B, manganese monitored at 257.6 nm. (Reproduced by permission from S.A. Estes, P.C. Uden, M.D. Rausch and R.M. Barnes, *Journal of High Resolution Chromatography and Chromatographic Communications*, 1980, 3, 471; copyright 1980 Dr. Alfred Huthig Publishers.)

excellent detection of iron, cobalt, nickel, chromium, manganese and vanadium in a series of cyclopentadienyl carbonyl/nitrosyl compounds, verifying elution of some previously unchromatographed compounds [45]. In Fig. 1.9 are seen chromatograms with carbon, chromium and manganese element specific detection. The identities of the eluted peaks in the "universal" carbon mode (left) are (a) $C_5H_5Mn(CO)_3$; (b) $CH_3C_5H_4Mn(CO)_3$; (c) $C_5H_5Cr(NO)(CO)_2$ and $(C_5H_5)_2Ni$ (unresolved); (d) $C_5H_5V(CO)_4$; (e) $(C_5H_5)_2Fe$ and (f) $C_5H_5(CH_3)_5Co(CO)_2$. On the right are shown manganese and chromium specific chromatograms. The slight responses noted for other compounds in these latter traces

18

is due to incomplete selectivity among the different metals at the emission lines measured, rather than any indication of the presence of these metals in other components of the mixture.

It may be noted that the manganese organometallics detected have been measured as gasoline additives by GC-AED using a DC Plasma system [46] and GC-MIP could undoubtedly be used in a similar way provided that solvent removal is accomplished following sample injection.

An interesting example of trace metal GC-MIP analysis, utilizing derivatization for organic functional group determination, involves the reaction of ferroceneboronic acid with diols. Brooks and Cole showed quantitative derivatization to give products with good GC behavior [47]. Figure 1.10 shows chromatographic separation of pinacol and dicyclohexyl-1,1'-diol ferrocene boronates measured by iron and by boron specific detection. The slight tailing noted on the second boron peak is due to plasma and not chromatographic effects, since it does not appear in the iron trace. It is probably due to deposition of boron on the plasma tube walls and may be minimized by doping hydrogen at ca. 1 ml min^{-1} into the plasma. The ability to carry out dual element quantitation gives an enhanced degree of credibility to analytical results [48].

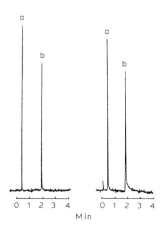

Fig. 1.10. Dual sequential GC-AED detection of (a) pinacol ferroceneboronate and (b) dicyclohexyl-1-1'-diol ferroceneboronate with left, iron monitoring at 259.94 nm and right, boron monitoring at 249.68 nm. Capillary FSOT column 12 m × 0.32 mm i.d. DB-5, temperature programmed from 210 to 260°C at 30°C min^{-1}. (Reproduced by permission from P.C. Uden, Y. Yoo, T. Wang and Z. Cheng, *J. Chromatogr.*, 1989, 468, 319; copyright 1989, Elsevier Science Publishers.)

1.5. FUTURE DIRECTIONS FOR CHROMATOGRAPHIC PLASMA EMISSION DETECTION

In real terms, wide adoption of plasma spectral detection will depend on the availability of commercial instrumentation to allow interlaboratory comparison of information and development of "standard" methods of analysis which can be widely employed. Many analyses are subject to constraints aimed at ensuring high levels of accuracy and precision. Plasma chromatographic detection has already demonstrated a wide utility and the recent commercial introduction of an integrated GC-MIP system suggests that the future of this technique is sound, despite earlier setbacks. Fully integrated units which circumvent the need for analysts to interface their own chromatograph, emission device and spectrometer may become as familiar in the future as GC-MS and GC-FTIR systems are today.

REFERENCES

1   P.C. Uden, J. Chrom., 313 (1984) 3–31.
2   L. Ebdon, S. Hill and R.W. Ward, Analyst, 111 (1986) 1113–1138.
3   L. Ebdon, S. Hill and R.W. Ward, Analyst, 112 (1987) 1–16.
4   P.C. Uden, Trends in Anal. Chem., 6(9) (1987) 238–246.
5   M. Zerezghi, K.J. Mulligan, and J.A. Caruso, J. Chromatog. Sci. 22 (1984) 348–352.
6   W.R. McLean, D.L. Stanton and G.E. Penketh, Analyst, 98 (1973) 432–442.
7   K.S. Brenner, J. Chromatog., 167 (1978) 365–380.
8   K.J. Slatkavitz, L.D. Hoey, P.C. Uden and R.M. Barnes, Anal. Chem., 57 (1985) 1846–1851.
9   H.J. Perpall, P.C. Uden and R.L. Deming, Spectrochim. Acta, 42B (1987) 243–251.
10 Y. Takigawa, T. Hanai and J. Hubert, J. High. Res. Chrom & Chrom. Commun., 9 (1986) 698–702.
11 R.L. Firor, American Lab., 21(5) (1989) 40–48.
12 A.J. McCormack, S.C. Tong and W.D. Cooke, Anal. Chem., 37 (1965) 1470–1476.
13 C.A. Bache and D.J. Lisk, Anal. Chem., 39 (1967) 786–789.
14 T.H. Risby and Y. Talmi, CRC Crit. Rev. in Anal. Chem., 14(3) (1983) 231–265.
15 F.C. Fehsenfeld, K.M. Evenson and H.P. Broida, Rev. Sci. Instr., 36(3) (1965) 294–298.
16 C.I.M. Beenakker, Spectrochim. Acta, 31B (1976) 483–486.
17 M.H. Abdellah, S. Coulombe and J.M. Mermet, Spectrochim. Acta 37B, (1982) 583–592.
18 W.J. Hoskin, "The MPD 850 Organic Analyzer", Applied Chromatography Systems, Ltd, Luton, U.K. 1977.
19 C. Bradley and J.W. Carnahan, Anal. Chem., 60 (1988) 858–863.
20 A. Bollo-Kamara and E.G. Codding, Spectrochim. Acta 36B, (1981) 973–982.
21 S.R. Goode, B. Chambers and N.P. Buddin, Spectrochim. Acta. 40B, (1985) 329–333.

20

22 S.A. Estes, P.C. Uden and R.M. Barnes, Anal. Chem., 53 (1981) 1336–1340.
23 H.A.H. Billiet, J.P.J. van Dalen, P.J. Schoemakers and L. deGalen, Anal. Chem., 55 (1983) 847–851.
24 D. Kollotzek, D. Oechsle, G. Kaiser, P. Tschopel and G. Tolg, Fresenius Z. Anal. Chem., 318 (1984) 485–489.
25 D.L. Haas, J.W. Carnahan and J.A. Caruso, Appl. Spectroscopy, 37 (1983) 82–89.
26 K.E. LeFreniere, V.A. Fassel and D.E. Eckel, Anal. Chem., 59 (1986) 879–887.
27 L. Zhang, J.W. Carnahan, R.E. Winans and P.H. Neill, Anal. Chem., 61 (1989) 895–897.
28 D.R. Luffer, L.J. Galante, P.A. David, M. Novotny and G.M. Hieftje, Anal. Chem., 60 (1988) 1365–1369.
29 C.B. Motley, M. Ashraf-Khorassani and G.L. Long, Applied Spec. 43, (1989) 737–741.
30 R.M. Dagnell, T.S. West and P. Whitehead, Analyst, 98 (1973) 647–654.
31 M.S. Black and R.E. Sievers, Anal. Chem., 48 (1976) 1872–1874.
32 Y. Talmi and V.E. Norvall, Anal. Chem., 47 (1975) 1510–1516.
33 Y. Talmi and D.T. Bostick, Anal. Chem., 47 (1975) 2145–2150.
34 D.C. Reamer, W.H. Zoller and T.C. O'Haver, Anal. Chem., 50 (1978) 1449–1453.
35 D.S. Sklarew, K.B. Olsen and J.C. Evans, Chromatographia, 27 (1989) 44–48.
36 K.B. Olsen, D.S. Sklarew and J.C. Evans, Spectrochim. Acta, 40B (1985) 357–365.
37 S.A. Estes, P.C. Uden and R.M. Barnes, Anal. Chem., 53 (1981) 1829–1837.
38 S.A. Estes, P.C. Uden and R.M. Barnes, Anal. Chem., 54 (1982) 2402–2405.
39 M.A. Eckhoff, J.P. McCarthy and J.A. Caruso, Anal. Chem., 54 (1982) 165–168.
40 R.W. Moshier and R.E. Sievers, "Gas Chromatography of Metal Chelates", Pergamon Press, New York, 1965.
41 P.C. Uden and D.E. Henderson, Analyst, 102 (1977) 889–916.
42 T.R. Crompton, "Gas Chromatography of Organometallic Compounds", Plenum Press, New York, 1982.
43 P.C. Uden, in "Inorganic Chromatographic Analysis". J. MacDonald. ed. John Wiley and Sons, New York, 1985 pp. 229–284.
44 P.C. Uden and T. Wang, J. Anal. Atomic Spec., 3 (1988) 919–922.
45 S.A. Estes, P.C. Uden, M.D. Rausch and R.M. Barnes, J. High Res. Chrom., 3(9) (1980) 471–472.
46 P.C. Uden, R.M. Barnes and F.P. DiSanzo, Anal. Chem., 50 (1978) 852–855.
47 C.J.W. Brooks and W.J. Cole, J. Chromatogr., 339 (1987) 207–217.
48 P.C. Uden, Y. Yoo, T. Wang and Z. Cheng, J. Chromatogr., 468 (1989) 419–428.

I.S. Krull (Ed.), *Trace Metal Analysis and Speciation*
Journal of Chromatography Library Series, Vol. 47
© 1991 Elsevier Science Publishers B.V., Amsterdam

Chapter 2

Determination of Organometallic Compounds in Environmental Samples with Element-Specific Detectors

KURT J. IRGOLIC
Department of Chemistry, Texas A&M University, College Station, TX 77843, U.S.A.

2.1. INTRODUCTION

Concern about environment effects of chemicals and progress in analytical chemistry impelled many investigators to analyze samples of air, water, soil, rocks, plants, animal organs, human tissues, and human fluids for metals and metalloids. Better analytical instrumentation, improvement in methods for the preparation of samples, and increased awareness of the effects of interferences on the results of analyses, moved the detection limits of the procedures for the determination of elements from the mg l^{-1} (mg kg^{-1}) range to μg l^{-1} (kg) and in some cases to ng l^{-1} (kg) levels. The data that have been accumulating over the past three decades indicate clearly, that human activities have accelerated the dispersion of metals and metalloids into the environment [1] and changed global chemical cycles [2]. The mobilization of metals, particularly of heavy metals, caused their concentrations in water and soils to increase. Plants—in turn—had higher levels of these metals in their tissues and passed them up the food chain [3]. Acute and chronic poisoning episodes caused by heavy metals such as lead, cadmium, mercury, arsenic, and selenium, are well documented [4]. To protect plants, animals, and man from excessive intakes of health-threatening metals, the daily dose associated with an acceptable low risk must be established. Suggestions to lower such doses to "zero" cannot be implemented. A "zero" dose or concentration can only be defined in terms of the state-of-the-art of analytical chemistry. A more serious objection to this concept of "zero" is the likelihood, that very low concentrations of metals in the diet might lead to deficiency diseases. History has clearly shown that elements considered to be without value to biochemical processes, were discovered on closer examination to be essential for life. Selenium has experienced such a change, and arsenic is on the way to its recognition as a life-supporting element [5].

Regulatory decisions about allowable and safe concentrations of elements in air, water, and food items are almost always based on "total element" concentrations. However, toxicologists and nutritional experts are well aware of the fact, that the undesirable inimical and the welcome beneficial effects of an element are caused not by the "element" but by specific compounds containing the element. An illuminating example is the much maligned element arsenic, of which no more than 50 micrograms are allowed to be present in a liter of U.S. drinking water. Almost all of the arsenic in drinking water is present as arsenate, an inorganic pentavalent arsenic compound. Seafood, harvested from the ocean that has an average arsenic (arsenite and arsenate) concentration of a few micrograms per liter saltwater, contains frequently several milligrams of arsenic per kilogram of tissue. These arsenic concentrations are approximately 500-times higher than the acceptable concentrations in drinking water. On the basis of total arsenic concentrations and the known toxic effects of inorganic arsenic, consumers of seafood might find reasons to fear for their health. Fortunately, most of the arsenic in seafood is in an organic form (arsenobetaine, carboxymethyltrimethylarsonium zwitterion) that was found to be innocuous [6]. Setting limits for arsenic in seafood in terms of "total" arsenic at levels reasonable for drinking water would not be very sensible, because this "marine arsenic" is not harmful. Enforcement of a 50 microgram arsenic per kilogram seafood limit would destroy most of the fishing industry.

Knowledge about the chemical forms in which elements occur in the environment, about the manner in which these species interact with organisms, and about the chemical changes these compounds experience in their journey through the food web is of utmost importance for environmental scientists, for toxicologists, nutritionists, practitioners of the healing arts, and for regulators [7]. The identification and quantification of trace element compounds is a more complex task than the determination of "total" element concentrations. However, methods have been developed for this more complex task that allow chemical species of many elements to be identified and quantified. These analyses are not routine as total element determinations are nowadays, but can be carried out by skilled analysts in possession of the required instrumentation. In the following sections, the state-of-the-art of trace-element-compound determinations using element-specific detectors will be presented. Arsenic compounds will be used frequently as examples,

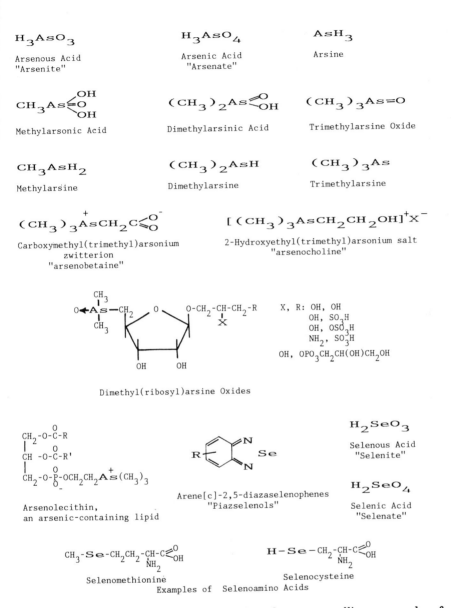

H_3AsO_3

Arsenous Acid
"Arsenite"

H_3AsO_4

Arsenic Acid
"Arsenate"

AsH_3

Arsine

$CH_3As{<}^O_{OH}^{OH}$

Methylarsonic Acid

$(CH_3)_2As{<}^O_{OH}$

Dimethylarsinic Acid

$(CH_3)_3As=O$

Trimethylarsine Oxide

CH_3AsH_2

Methylarsine

$(CH_3)_2AsH$

Dimethylarsine

$(CH_3)_3As$

Trimethylarsine

$(CH_3)_3\overset{+}{A}sCH_2C{<}^{O^-}_O$

Carboxymethyl(trimethyl)arsonium
zwitterion
"arsenobetaine"

$[(CH_3)_3\overset{+}{A}sCH_2CH_2OH]^+X^-$

2-Hydroxyethyl(trimethyl)arsonium salt
"arsenocholine"

X, R: OH, OH
OH, SO$_3$H
OH, OSO$_3$H
NH$_2$, SO$_3$H
OH, OPO$_3$CH$_2$CH(OH)CH$_2$OH

Dimethyl(ribosyl)arsine Oxides

Arsenolecithin,
an arsenic-containing lipid

Arene[c]-2,5-diazaselenophenes
"Piazselenols"

H_2SeO_3

Selenous Acid
"Selenite"

H_2SeO_4

Selenic Acid
"Selenate"

Selenomethionine

Selenocysteine

Examples of Selenoamino Acids

Fig. 2.1. Formulas and names for inorganic and organometallic compounds referred to in the text of this chapter.

because of our extensive and intensive experience with the environmental chemistry of arsenic. These arsenic compounds and other compounds mentioned in this chapter are identified by name and structure in Fig. 2.1.

References pp. 45–48

2.2. THE TERM "SPECIATION"

In publications reporting work on trace element determinations, the term "speciation" is frequently used. This term—borrowed from biology—is found to have several meanings, often within the same paper. This practice is against the principle that a scientific term shall be well defined and shall be used unambiguously only in one meaning. "Speciation" has been used to designate the assembly of trace element compounds in a sample (static sense), to express the idea that trace element compounds react and are transformed (kinetic sense), and to refer to the process of determining trace element compounds (operational sense). "Speciation" in the operational sense appears to be most useful at this time and should be used exclusively with this meaning. However, to avoid ambiguities and confusion, a good case can be made for abstaining from the incorporation of the term "speciation" into any paper. Concise, substitute terms, such as assembly of species, species distribution, transformation of species, and determination of trace element compounds, are available.

2.3. ADVANTAGES OF ELEMENT-SPECIFIC DETECTORS

"Benign" matrices resembling solutions of reagent-grade substances in distilled water do not cause many problems. For instance, an aqueous solution containing selenite, selenate, and sulfate at comparable concentrations and other ions can be easily analyzed by gradient ion chromatography/conductometric detection, with all signals well separated [8]. However, when the concentrations of selenite and selenate are small in comparison to sulfate, "general" detectors that do not respond specifically to selenium will be of little use. An ion chromatogram of a synthetic river water containing per liter 277 mg chloride, 69 mg sulfate, 5 mg phosphate, 0.4 mg selenium as selenite, and 0.4 mg selenium as selenate is shown in Fig. 2.2.

The conductivity detector was set at a sensitivity that would have produced an acceptable signal for the selenium compounds in the absence of the other ions. Because of the high sensitivity setting on the detector and the high concentrations of chloride, sulfate, and phosphate relative to selenite and selenate, the recorder pen is off the chart for retention times between 8 and 25 minutes, and signals for selenite and selenate cannot be reliably located. However, when a graphite furnace atomic absorption spectrometer was employed as the selenium-specific

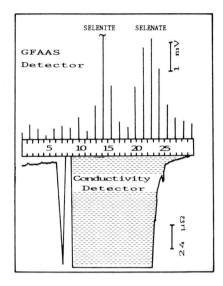

Fig. 2.2. Ion chromatograms of a synthetic river water (277 mg l^{-1} chloride, 69 mg l^{-1} sulfate, 5 mg l^{-1} phosphate) spiked with 400 micrograms selenium as selenite and 400 micrograms selenium as selenate recorded with a conductivity detector and a Hitachi Zeeman graphite furnace atomic absorption spectrometer (GFAAS) as the selenium-specific detector (Dionex Model 16 ion chromatograph, 1.0 ml sample, 50 × 3 mm anion precolumn Dionex 30008; mobile phase 0.008 M aqueous Na_2CO_3, 0.46 ml min^{-1}; 150 × 3 mm anion separator column Dionex 30589; 250 × 3 mm anion suppressor column Dionex 30066. GFAAS: drying 120°, 60 sec; no ashing; atomization 2500°, 6 sec; Se lamp 10 mA, 196.0 nm; 80 sec between injections; retention time in min). Redrawn from the *Journal of Chromatography* [9] by permission of Elsevier Science Publishers and the authors.

detector, the chromatogram consisted of two sets of signals, one set each for selenite and selenate [9]. Sulfate, chloride, and phosphate did not interfere with the detection of the selenium compounds. Even concentrations as high as 800 mg l^{-1} of sulfate did not impair the detection and quantification of selenite and selenate at 0.4 mg Se per liter. Only a slight peak broadening was observed for selenate with increasing sulfate concentration [9].

High pressure liquid chromatography is frequently used to separate and identify trace element compounds. A common detector for liquid chromatographs is the differential refractive index detector that is not element-specific. To achieve separation, mobile phase gradients or sudden changes in the composition of mobile phases are frequently needed. The differential refractive index detector will often not be able to compensate for changes in the refractive index of the mobile phase and,

References pp. 45–48

26

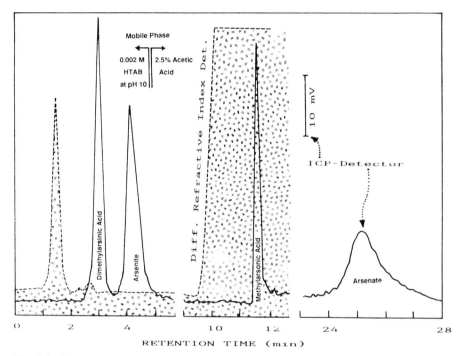

Fig. 2.3. Chromatogram of a solution containing arsenite, arsenate, methylarsonic acid and dimethylarsinic acid (each 100 mg As l^{-1}) recorded with a differential refractive index detector and an ARL 34000 simultaneous inductively coupled argon plasma atomic emission spectrometer as the arsenic-specifier detector [Hamilton PRP-1 column, hexadecyltrimethylammonium bromide (HTAB) as ion-pairing reagent; flow rate 1.5 ml min^{-1}; 0.1 ml injected; Waters Associates high pressure liquid chromatograph; integration time on ICP 5 sec, As 189.0 nm]. Redrawn from *Marine Chemistry* [10] by permission of Elsevier Science Publishers and the authors.

therefore, produce signals that may be interpreted erroneously as arising from a trace element compound. The signals caused by the change in mobile phase may be so intense as to obscure signals from trace element compounds. An example of total obscuration of signals is shown in Fig. 2.3. A series of arsenic compounds was separated by reverse-phase high pressure liquid chromatography. The mobile phase was changed five minutes after the start of the chromatographic analyses from a 0.002 M aqueous solution of hexadecyltrimethylammonium bromide to 2.5% aqueous acetic acid. The differential refractive index detector could not compensate for the change in refractive index of the mobile phase and signals for methylarsonic acid and arsenate did not appear in the chromatogram. However, a simultaneous, inductively coupled argon plasma emission spectrometer as the arsenic-specific detector was not affected by the change in mobile phase and clearly indicated the presence of methylarsonic acid and arsenate in the sample [10].

The use of element-specific detectors for chromatography offers an additional bonus. Whereas with a general detector all the compounds one wants to identify and quantify in a sample must be reasonably separated from each other, only compounds with the same element in the molecule must be separated when an element-specific detector is used. Thus, arsenate does not have to be separated chromatographically from phosphate, which is chemically very similar to arsenate, when either an arsenic-specific or a phosphorus-specific detector is coupled to the chromatograph. Even greater advantages can be enjoyed, when the detector is multi-element-specific as simultaneous, inductively coupled argon plasma emission spectrometers, inductively coupled argon plasma-mass spectrometers, and direct-current plasma emission spectrometers are. These advantages are pictured (Fig. 2.4) in a high pressure liquid chromatogram of a solution containing five arsenic compounds, phosphate, and selenite obtained with a reverse-phase column and a simultaneous, inductively coupled argon plasma emission spectrometer monitoring arsenic, phosphorus, and selenium emissions. Phosphate and arsenate have exactly the same retention time. A general detector could neither provide information about the nature of the oxoanion producing the signal nor the concentration ratio of the two oxo-species. The selenite signal in a general-detector trace would appear as a shoulder on the low-retention-time side of the arsenate/phosphate peak. The plasma emission spectrometer monitors the signals separately and displays the signal for each element in a separate trace [11].

As a rule, environmental samples, in which trace element compounds are to be determined, are rather complex. Soil extracts, porewaters, extracts from algae, rice leaves, shark livers, mushrooms, and crude petroleum, and body fluids of human or animal origin are very far in composition from solutions of standard compounds in distilled water. Most of these samples could be characterized as thick soups of unwanted neutral and charged, organic and inorganic compounds and ions, with a sprinkling of the trace element compounds of interest. A detector that is not element-specific faces almost insurmountable problems, unless the trace element compounds are cleanly separated from all other components in the sample. At the sensitivity settings required to obtain sufficiently intense signals for trace element compounds, the peaks for the higher-concentration constituents will become very intense and broad. The chromatography has to carry the heavy burden of separating

28

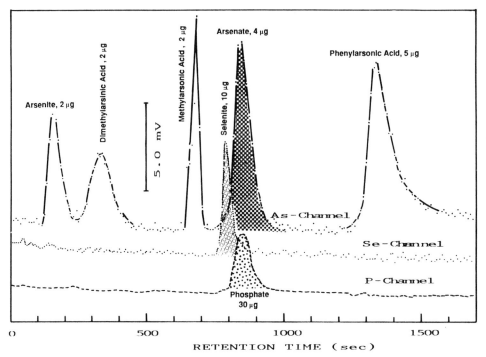

Fig. 2.4. Chromatogram of a solution containing arsenite, arsenate, methylarsonic acid, dimethylarsinic acid, phenylarsonic acid, selenite, and phosphate recorded with an ARL 34000 simultaneous inductively coupled argon plasma emission spectrometer as the multi-element-specific detector [Hamilton PRP-1 resin-based reverse-phase column, Waters Associates Inc. Model 6000A high pressure liquid chromatograph, 0.1 ml injected; flow rate 1.5 ml min^{-1}, mobile phases: 0.002 M aqueous HTAB at pH 9.6 to 250 sec, 99/1 (v/v) H_2O/CH_3COOH 250–1100 sec, 90/10 (v/v) H_2O/dimethylformamide 1100–1700 sec. ICP As 189.0 nm, P 241.9 nm, Se 203.9 nm, integration time 5 sec]. Redrawn from *Spectrochimica Acta* [11] by permission of Pergamon Press and the authors.

mixtures with components, the concentrations of which may differ by factors of several thousand. In general, the time required to obtain usable chromatograms will be rather long. Under these circumstances, the element-specific detectors show their advantages by simplifying the chromatography, circumventing the problems associated with the sensitivity settings on general detectors, and being silent towards all analytes not containing the element(s) chosen to be detected.

Element-specific detectors can and have been used in conjunction with gas chromatographs, with ion chromatographs, and with high pressure liquid chromatographs to great advantage. Most of the element-specific detectors are based on atomic absorption or emission spectrometry. Commercial spectrometers that are unfortunately not inexpensive can be coupled to chromatographs without much difficulty. Because graphite furnace atomic absorption spectrometers are available in most laboratories performing trace element analyses, the combination of chromatographs with such spectrometers is the easiest entry into element-specific systems suitable for the determination of trace element compounds.

2.4. DETERMINATION OF VOLATILE TRACE COMPOUNDS

Biological methylation is known to convert inorganic compounds of antimony, arsenic, mercury, selenium, tellurium, tin and perhaps lead to methylated derivatives [12]. The fully methylated compounds [$(CH_3)_3As$, $(CH_3)_3Sb$, $(CH_3)_2Hg$, $(CH_3)_2Se$, $(CH_3)_2Te$, $(CH_3)_4Sn$, $(CH_3)_4Pb$] and some partially methylated compounds [for instance, Ch_3AsH_2, $(CH_3)_2AsH$] have boiling points below 110°, possess considerable vapor pressures at room temperature, and are sufficiently stable toward atmospheric agents to appear in the gas phase around organisms with methylating capability. A laboratory worker, who has ingested traces of inorganic selenium compounds because of improper handling of these chemicals, is an example of such an organism. The dimethyl selenide formed in the body is excreted through the skin and is easily detected by the noses of the coworkers. The identification and quantification of volatile trace element compounds of organic or inorganic nature does not pose any problems. The entire array of analytical instruments and methods developed for volatile organic compounds, is in principle, available for the determination of these volatile organometallic compounds.

2.4.1. Atomic absorption detectors

Gas chromatographs equipped with suitable columns [13] and coupled to an element-specific detector can separate, identify, and quantify gaseous organometallic compounds. The first attempts of using flame atomic absorption spectrometers as metal-specific detectors for gas chromatography date to 1966 [14]. Flame systems were soon replaced by heated quartz cells [15] of various designs or by electrically heated graphite tubes [16, 17], in which the gaseous metal compounds were atomized and the metal atoms detected by absorption spectrometry.

References pp. 45–48

These early GC-element-specific detection systems are described in several review articles [18–21]. An excellent summary of the applications of these and other systems, to be discussed later, to the determination of mercury, lead, selenium, tin and arsenic compounds is provided in the review of this field by C.J. Cappon [9]. The detection limits obtained with these systems reach the low picogram levels under ideal conditions; however, detection limits in the nanogram range, are more common. With elemental mercury or organic mercury compounds in a sample, 100 picograms of mercury were detected in a system consisting of a fused silica capillary gas chromatograph and a cold vapor atomic absorption spectrometer [22].

2.4.2. Plasma emission detectors

The advent of plasma emission sources combined with polychromators provided opportunities to couple gas chromatographs with multi-element-specific detectors. Analytes introduced into helium or argon plasmas, the temperatures of which are much higher than encountered in flames and graphite furnaces employed in atomic absorption spectrometry, are efficiently atomized and the resulting atoms excited. The high plasma temperatures excite even atoms with high ionization potential and assure intense emissions from all the elements. Three types of plasmas are available: microwave-induced plasmas, direct-current plasmas, and inductively-coupled plasmas. Argon or helium serve in most cases as plasma gases.

2.4.2.1. Microwave-induced plasma detectors

The first GC-microwave-induced plasma emission system was reported in 1965 [23]. During the past two decades GC-plasma emission systems have gained in popularity and have been used for the identification and quantification of mercury, lead, tin, selenium, and arsenic compounds [13]. The most frequently used plasma source is the microwave-induced plasma operated either at reduced pressure or at atmospheric pressure with helium or argon as the plasma gases at powers of 100 to 200 W. The Beenakker cylindrical resonance cavity introduced in 1976 [24], and since then modified to achieve better detection limits, is most frequently used in the GC-microwave-induced plasma emission systems that are easily adaptable to capillary GC operation. These microwave-induced plasma detectors respond to non-metals (H, D, B, C, N, O, F, Si, P, S, Cl, As, Se, Br, I) and metals, with absolute detection limits in the low picogram range in favorable cases. The reduced-pressure microwave system has better detection limits for oxygen and nitrogen than the atmospheric plasmas; however, the atmospheric plasmas are easier to operate. A fully automated capillary gas chromatograph–atmospheric pressure microwave-induced plasma system operating with helium as the plasma gas and capable of detecting any element except helium is now available commercially. The spectrometer uses a flat focal planne with a movable photodiode array responding to light with wavelengths in the range 170 to 780 nanometers [25]. Several excellent reviews of GC-microwave-induced plasma systems are available in the literature [13, 26–28]. An interesting application involves the use of the boron emission for the determination of traces of catechols in form of their butylboronic esters [29] and for the estimation of steroidal carboranes [30].

2.4.2.2. *Direct-current plasma emission detectors*

The commercially available direct-current argon plasma spectrometer has also been interfaced with gas chromatographs for the detection of metals and non-metals. The DC systems are not as well suited for work with capillary columns as the microwave-induced plasma systems. However, DC plasmas are more robust than microwave plasmas and, therefore, can handle larger quantities of solvent or analyte. The absolute detection limits achieved with the GC-DC systems are often comparable with the detection limits in the microwave systems, but are two orders of magnitude higher for lead and mercury [28]. The DC plasma emission spectrometers have found little use for the determination of organometallic compounds after GC separation. Examples of such applications are the detection of copper, nickel, and palladium complexes [31], the detection of silicon in trimethylsilyl derivatives of phenolic acids in corn meal extracts [27], the determination of (methylcylopentadienyl)tricarbonylmanganese in gasoline [32], the detection of various sulfur compounds [33], and the determination of methyl mercury in fish [34].

2.4.2.3. *Inductively-coupled argon plasma emission detectors*

Inductively coupled argon plasma emission spectrometers of the sequential or simultaneous type were used very little as detectors for gas chromatographs [28]. The detection limits for metals in these systems approached the low nanogram levels. However, the detection limits for

References pp. 45–48

nonmetals were much inferior to the detection limits achieved with the GC-microwave-induced plasma systems [35]. The high cost of inductively coupled plasma emission spectrometers is an effective deterrent to their wide use as element-specific detectors for gas chromatographs.

2.4.3. Mass spectrometry detectors

The combination of a gas chromatograph with a mass spectrometer is ideal for the identification and quantification of gaseous organometallic compounds. Such systems, operating in many laboratories, are mainly dedicated to work with volatile organic compounds. The mass spectrometer serves not only as an element-specific detector, but also as a molecule-specific detector provided the analytes produce molecular ions under the analytical conditions. With organometallic compounds in a complex matrix, for instance in a matrix of hydrocarbon gases, the separation will be enhanced by low-temperature gas chromatography that would eliminate most of the matrix consisting of very low boiling gases. Even if the separation of organometallic compounds and matrix hydrocarbons is not complete, evidence for the presence of an organometallic compound can be obtained from the mass spectra, provided the mass spectrometer has a sufficiently high resolving power. Organometallic compounds containing a reasonably "heavy" element will produce mass spectrometric signals that can be distinguished from signals coming from ions consisting of light elements only. The "heavy-ion" signal will have an exact mass lower than the "light-ion" signal caused, for instance, by a hydrocarbon of the same nominal m/z value. This mass defect for arsenic-containing compounds amounts to approximately 0.1 mass units. Compared to the large volume of work with organic compounds, hardly any use has been made of the capabilities of gas chromatography-mass spectrometry for organometallic compounds in environmental samples.

2.5. DETERMINATION OF TRACE ELEMENTS AND TRACE ELEMENT COMPOUNDS THAT ARE CONVERTIBLE TO VOLATILE DERIVATIVES

Because only a small number of organometallic compounds have boiling points low enough to be found as gases at ambient temperatures, the methods for the identification and quantification of such compounds by gas chromatography-element-specific detection would have only limited applicability. Fortunately, a relatively large number of environmentally important molecules that contain metals and metalloids can be converted by derivatization to compounds possessing sufficient volatility for gas chromatographic separation. Examples of such derivatizations are reactions of mercury(II) cations with methylating agents to form methyl mercury chloride, the conversion of alkyl lead cations, $R_n Pb^{(4-n)+}$, with Grignard reagents to tetraalkyl lead compounds, the reduction and complexation of methylarsonic acid with diethyldithiocarbamate, and the cyclization of aromatic ortho-diamines with selenous acid to arene[c]-2,5-diazaselenophenes [13]. The compounds formed in these reactions can then be determined by gas chromatography (section 2.4). Frequently, the volatile derivatives are concentrated by extraction into an organic solvent prior to injection into the gas chromatograph.

2.5.1. The hydride generation method

One of the most widely used derivatizations is the reduction of non-volatile compounds to volatile hydrides [36]. This hydride generation method is—in principle—applicable to the determination of all elements that form volatile hydrides and that can be detected by an element-specific detector. When only inorganic compounds are present or when all organometallic compounds were converted to inorganic compounds by oxidative mineralization, reducing agents such as zinc in hydrochloric acid or sodium borohydride will convert the inorganic compounds into their hydrides. For instance, selenite will be reduced to hydrogen selenide, H_2Se, and arsenate to arsine, AsH_3. Inorganic mercury, lead, tin, antimony, and bismuth compounds are transformed similarly to hydrides. These hydrides are then flushed from the reaction mixture by an inert gas such as helium and passed into an appropriate element-specific detector. The detection limits of a hydride generation are dependent on the nature of the element to be determined, on the detector employed, and on additional devices incorporated into the system. Absolute detection limits (the quantity of an element that must reach the detector) in the low nanogram range are easily achieved. Detection limits thousand-times lower are also possible under appropriate conditions.

2.5.1.1. *Detectors for hydride generation systems*

A hydride generation system in its simplest form (Fig. 2.5) consists of the hydride generator, in which the conversion to the volatile hydrides

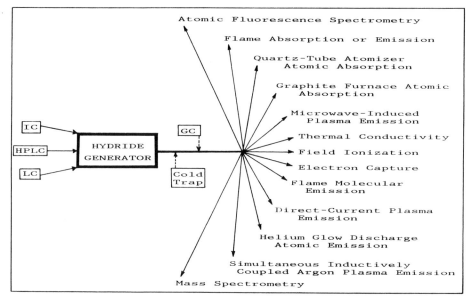

Fig. 2.5. Schematic representation of hydride generation systems for the identification and determination of hydride-forming trace elements and trace element compounds with element-specific and general detectors.

occurs, the transfer lines to the detector, and the detector. A wide choice of detectors is available.

Common gas chromatographic detectors that are not element- or metal-specific, atomic absorption and atomic emission detectors that are element-specific, and mass spectrometric detectors have all been used with the hydride systems. Flame atomic absorption and emission spectrometers do not have sufficiently low detection limits to be useful for trace element work. Atomic fluorescence [37] and molecular flame emission [38–40] were used by a few investigators only. The most frequently employed detectors are based on microwave-induced plasma emission, helium glow discharges, and quartz tube atomizers with atomic absorption spectrometers. A review of such systems as applied to the determination of arsenic, associated with an extensive bibliography, is available in the literature [36]. In addition, a continuous hydride generation system was coupled to a direct-current plasma emission spectrometer for the determination of arsenite, arsenate, and total arsenic in water and tuna fish samples [41].

Additional features can be added to the basic hydride generation system to make it more versatile. When more than one hydride-forming element is to be determined, a multi-element-specific detector is very useful, because all the hydrides can be passed, for instance, into a plasma with the emitted light from the various elements sorted by a polychromator. The plasma polychromator detectors suffer from the drawback of high acquisition cost. When single-element-specific detectors (atomic absorption spectrometer, plasma emission/monochromator combination) are associated with a hydride generator, then the element hydrides must be separated before reaching the detector, and the detector must be set to respond to the element to be detected. Such a separation can be achieved by placing a gas chromatograph equipped with an appropriate column between the hydride generator and the detector.

The performance of a hydride generation system consisting of a hydride generator and a detector, with or without a gas chromatograph, is influenced by the rates of formation of the hydrides in the hydride generator. To become independent of kinetic limitations, a cold trap cooled with liquid nitrogen is frequently placed just after the hydride generator. The hydrides are flushed with a stream of helium from the generator into the cold trap where they condense. When the reduction is complete, the cold trap is warmed and the hydrides are volatilized to be separated and detected. Additional traps may be used to remove from the gas stream water and carbon dioxide that may interfere with the detection [42]. These hydride generation systems with element-specific detectors have excellent detection limits and produce precise and accurate results when operated by skilled personnel. Hydride generation attachments are commercially available for atomic absorption spectrometers and for plasma emission spectrometers. Hydride generation systems with arsenic-specific detectors have found wide application for the identification and quantification of inorganic and organic arsenic compounds that are reducible to volatile arsines. Examples of arsenic compounds convertible to arsines by sodium borohydride are: arsenite and arsenate (reduced to AsH_3), methylarsonic acid (reduced to CH_3AsH_2), dimethylarsinic acid [reduced to $(CH_3)_2AsH$], and trimethylarsine oxide [reduced to $(CH_3)_3AS$]. Arsenite is the only arsenic compound reduced at a pH of 5. All other reducible arsenic compounds including arsenite are converted to the respective arsines in an aqueous solution of approximately pH 1. This difference in reducibility allows arsenite to be determined separately and removed from the solution. The various arsines formed at pH 1 are collected in the cold trap and then volatilized in the order of increasing boiling point into the detector. Better separation of the arsines can be achieved with a gas chromatograph.

References pp. 45–48

2.5.1.2. *Liquid chromatographs and hydride generation systems*

The separation of reducible compounds can be accomplished by column liquid chromatography, ion chromatography, or high pressure liquid chromatography before the reduction step. The column effluent is then passed into the hydride generator, and the arsines flushed into the detector [37]. A trap for the collection of the arsines and a gas chromatograph are not needed in this continuous-flow system. Peristaltic pumps deliver and dose the reducing agent, most commonly an aqueous solution of sodium borohydride made basic by addition of sodium hydroxide. Mixing coils assure complete mixing of the analytes with the reducing agent and quick reduction, and gas–liquid separators send the gaseous arsines to the arsenic-specific detector and the liquids to waste. If necessary, the analytes can be concentrated on a suitable column before separation by liquid chromatography and reduction and detection in a hydride generation system [37]. Arsenite and arsenate were determined in water samples after separation on C_{18} columns by high pressure liquid chromatography, reduction to arsine and detection with an inductively coupled plasma emission spectrometer [43]. Methyltin compounds were identified and quantified similarly with a direct-current plasma emission spectrometer as the tin-specific detector [44].

2.5.1.3. *Hydride generation system with a mass spectrometric detector*

One of the to-date most sophisticated hydride generation systems was assembled by T. Kaise and coworkers for the identification of inorganic and organic arsenic compounds in marine organisms [45]. Their automated system uses a Hitachi Model HFS-2 hydride generator. Pumps deliver sodium borohydride solution, hydrochloric acid, water, and an aliquot of the sample to a mixer. The reaction mixture flows through a mixing coil into a gas–liquid separator. The separated gases diluted with helium as the carrier gas are passed through a liquid-nitrogen trap, where all the arsines are collected. The trap is constructed in the form of a "U" from a stainless steel tube, is packed with quartz wool, wrapped with Nichrome wire that is connected to a variable transformer, and insulated with an asbestos ribbon. When the reduction is complete and all the arsines are condensed in the trap, the coolant is removed and the U-tube heated electrically to 200°C to transfer the arsines into the gas chromatograph equipped with a 3-meter (3 mm i.d.) glass column packed with 3% silicone OV-17 on 80/100 mesh

Chromosorb W (AW, DMCS). The arsines are separated on this column and passed into a mass spectrometer operated in the electron impact mode (70 eV), with the ion source at 180°C. Ions with m/z of 76 and 78 for AsH_3, 90 for CH_3AsH_2, 90 for $(CH_3)_2AsH$, and 103 and 120 for $(CH_3)_3As$ are selectively monitored. The peaks at these m/z values are the most intense in the mass spectra of the arsines. The mass spectrometer used in the selective-ion monitoring mode serves as an element- and molecule-specific detector. This system allows arsenite to be determined by pH-controlled reduction (pH ~ 5). Total inorganic and reducible methylarsenic compounds are reduced at pH 1 (eqn. 1):

Sodium Borohydride Reduction (1)

at pH 5		of		at pH 1	
AsH_3	←	H_3AsO_3	→	AsH_3	bp. −56°
		H_3AsO_4	→	AsH_3	
		$CH_3AsO_3H_2$	→	CH_3AsH_2	bp. + 2°
		$(CH_3)_2AsO_2H$	→	$(CH_3)_2AsH$	bp. +36°
		$(CH_3)_3AsO$	→	$(CH_3)_3As$	bp. +53°

Marine organisms are known to contain, in addition to inorganic arsenic compounds and simple methylated arsenic compounds, the more complex arsenic derivatives arsenobetaine and ribosyldimethylarsine oxides [46, 47]. These two complex derivatives are converted upon treatment with 2 M aqueous sodium hydroxide for three hours at 95°C to reducible arsenic compounds, arsenobetaine to trimethylarsine oxide (eqn. 2) and ribosyldimethylarsine oxides probably to dimethylarsinic acid (eqn. 3):

$$(CH_3)_3\overset{+}{As}-CH_2COO^- \xrightarrow{NaOH} (CH_3)_3AsO \quad (2)$$

$$\xrightarrow{NaOH} (CH_3)_2AsO_2H \quad (3)$$

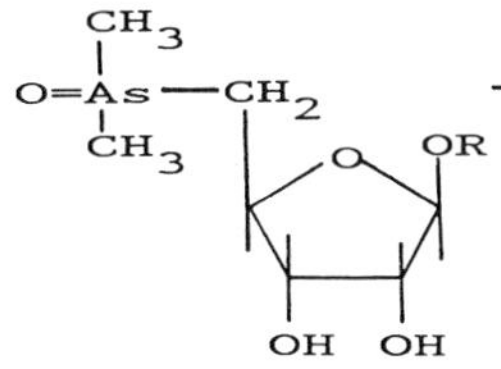

An aliquot of a sample without digestion with base analyzed with this hydride generation system will provide information about the presence of inorganic arsenic compounds and simple methylated arsenic compounds. Analysis of another aliquot that has been treated with sodium hydroxide will produce a more intense signal for trimethylarsine if arsenobetaine was present and for dimethylarsine if a ribosylarsine oxide was present. The intensity differences between the dimethylarsine signals are related to the ribosylarsine oxide concentration in the sample, and the differences between the trimethylarsine signals to the concentration of arsenobetaine [45].

The hydride generation method in its many variations [48], aided by a wide choice of element-specific detectors, makes it possible to identify and quantify many trace elements and trace element compounds in environmental samples with complex matrices. The detection limits achieved with these systems are, in many cases, the best available. The conversion to volatile hydrides has the great advantage of removing the analytes from the matrix and concentrating them before detection. The atomic absorption and emission detectors add specificity to the process. In spite of all these advantages, the hydride generation method is not free from interferences. Many reports in the literature address the influence of mineral acids on the reduction step, and the effects of the sample volume, nature of reducing agent, concentrations of anions, and concentrations of cations on the signal intensity. Interferences might occur during reduction, during transport of the hydrides, and in the detector. A review of such interferences with the determination of arsenic is available in the literature [36]. The hydride generation method must always be used with appropriate consideration of potential interferences from the matrix.

2.6. DETERMINATION OF TRACE ELEMENT COMPOUNDS NOT REDUCIBLE TO VOLATILE DERIVATIVES

Many trace element compounds are not volatile and cannot be converted to volatile derivatives without loss of information about their original chemical nature. These recalcitrant compounds cannot be identified and determined by the methods discussed in the earlier sections. Liquid chromatography, particularly ion chromatography and high pressure liquid chromatography, will make it possible to separate such non-volatile trace element compounds. Selenium-containing proteins, selenoamino acids, arsenocholine, arsenic-containing lipids, methylcobalamin, and platinum-containing anticancer drugs are examples of such compounds. A large variety of stationary phases stable over a wide pH range, with an even larger variety of associated surfaces ranging from ionic to non-polar, are at the analyst's disposal. Element-specific detectors again simplify the analysis and generally improve the detection limits. Atomic spectroscopies allow element-specific detection in the column effluents. Microwave-induced plasmas, so useful as components of gas chromatograph/element-specific detection systems, are not robust enough to serve in liquid chromatography systems. Introduction of the liquid stream emanating from the column into the plasma will extinguish the plasma [49]. The use of microbore columns that operate with much smaller volumes of mobile phases than conventional HPLC columns may allow microwave-induced plasmas to become valuable detectors in liquid chromatography. Flame atomic absorption and emission spectrometers generally have detection limits that are too high for trace element determinations. Therefore, graphite furnace atomic absorption spectrometers, direct-current plasma emission spectrometers, and inductively coupled argon plasma atomic emission spectrometers have most often been used as single- or multi-element-specific detectors with liquid chromatographs [26–28, 36, 50–53].

2.6.1. Graphite furnace atomic absorption spectrometers

Graphic furnace atomic absorption spectrometers have excellent detection limits for metals, but cannot be used for the determination of most non-metals. The coupling of the continuously operating liquid chromatograph to the graphite furnace that cannot accept a continuous stream of liquid poses a problem. The graphite furnace must cycle through the drying, ashing and atomization steps and must cool close to room temperature before the next aliquot taken from the column effluent can be injected. The "low-technology" interface between the chromatograph and the furnace is a fraction collector that collects small volumes of the effluent and a technician who manually injects an aliquot of each collected fraction into the graphite furnace. This system allows any laboratory with a graphite furnace atomic absorption spectrometer and a fraction collector to determine trace element compounds by liquid chromatography with a single-element-specific detector. The resolution of such a system depends on the volume of the fractions collected and is not very high.

Brinckman and coworkers [54] introduced an automated interface connecting a high pressure liquid chromatograph to a graphite furnace atomic absorption spectrometer. A specially designed "well sampler" is placed into the appropriate location of a stopped autosampler carousel. The column effluent enters the well sampler from the bottom and leaves through the side duct connected to a water aspirator. The autosampler

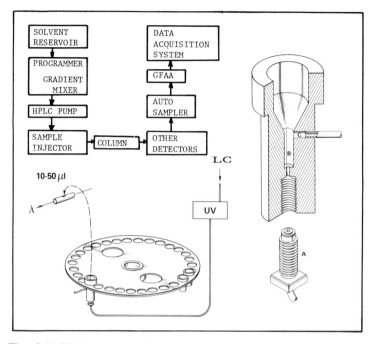

Fig. 2.6. High pressure liquid chromatography-graphite furnace atomic absorption spectrometer system with "well sampler" and autosampler as interface [A: dead volume screw; B: sample well; C: side duct connected to aspirator for withdrawal of liquid]. Redrawn from the *Journal of Chromatographic Science* [54] by permission of Preston Publications, a division of Preston Industries, Inc., and the authors.

arm dips into the well sampler and transfers an aliquot of the column effluent into the graphite furnace (Fig. 2.6). A chromatographic band is visualized in this manner as a series of signals, the intensities of which are proportional to the element-concentrations in the aliquots automatically injected into the graphite furnace (Fig. 2.2). The amount of analyte in a band can be obtained by summing the intensities of all the signals defining a band or by measuring the area of the peak defined by connecting the signal maxima with a smooth curve.

Without an autosampler, an interface can be constructed from an eight-port, two-position sample valve with sample and by-pass loops of appropriate volumes, an injection device that delivers aliquots of the column effluent isolated in the sample loop into the graphite furnace, and electronic control and switching circuitry that allows all steps in the analysis to occur in the correct sequence [55, 56]. All the components of the interface are commercially available. The injection device must be fitted to the spatial requirements of each type of graphite furnace. These systems can be assembled with or without a computer as the control and data collection device. Because of the requirement that the furnace must have sufficiently cooled to allow the introduction of the next liquid aliquot without splattering and the time needed for the drying, ashing, and atomization steps, the period between two analyses cannot be shorter than 30 seconds and is generally set at 45 to 60 seconds. Inorganic mercury, lead, tin, selenium, and arsenic compounds were determined with these systems [52, 53].

2.6.2. Direct-current and inductively coupled plasma emission spectrometers

The commercially available (but rather expensive) plasma emission spectrometers are much easier to connect to a liquid chromatograph than a graphite furnace atomic absorption spectrometer. All that is necessary is a plastic tube to join the column exit to the nebulizer. The flow rate through the column should match the uptake rate of the nebulizer. If the flow rate is too high, the liquid stream can be split. Direct-current and inductively coupled plasma emission spectrometers are designed for simultaneous determinations of a variety of elements and, therefore, may serve as multi-element-specific detectors. Whether a plasma emission spectrometer can be operated in this mode, depends on the operating software. In general, the software needs to be modified to successfully operate the spectrometer in the multi-element mode as a liquid chromatographic detector. The modifications are not involved provided listings of the programs are available. An example of such a modification and of an additional program allowing simultaneous on-line printout of five chromatograms with choice of elements is available in the literature [11]. Direct-current plasma emission spectrometers found surprisingly little use as detectors for liquid chromatography. Examples of such uses are the separation and determination of chromium(III) and chromium(VI) in environmental samples [57], of diethyldithiocarbamates of copper(II), nickel(II) and cobalt(II) [58], and of trifluoroacetylacetonates of chromium and cobalt. The simultaneous inductively coupled argon plasma spectrometer has been employed quite frequently as a single-element-specific detector for liquid chromatography [51–53], particularly for the identification and quantification of inorganic and organic arsenic compounds in water samples and in extracts from marine organisms. Additionally, organic mercury compounds [60, 61], metal-containing proteins [62], organically bound species of calcium, copper, iron, magnesium, titanium, and zinc in fractions from solvent-refined coal [63], lead compounds in gasoline [60], chromium(III) and chromium(VI) [64], arsenite and arsenate [65], and several metal cations [65] were investigated with these systems.

2.6.3. Comparison of graphite furnace atomic absorption and inductively coupled argon plasma emission spectrometers as element-specific detectors for liquid chromatography

No one system can claim to have all the advantages. The graphite furnace atomic absorption spectrometers are generally present in laboratories performing trace element analyses and can be adapted to serve as single-element-specific detectors for a liquid chromatograph when an autosampler is available. Without an autosampler, an interface must be constructed requiring some mechanical and electronic expertise. The graphite furnace has the advantage that almost all mobile phases of aqueous or organic nature can be used in the chromatography. During the drying step the mobile phase will be evaporated. The disadvantage comes with the compound-dependent response of the graphite furnace spectrometer. For instance, quantities of methylarsonic acid and dimethylarsinic with the same amount of arsenic injected into the graphite furnace, did not produce the same signal intensity [66]. This unfortunate fact makes quantification difficult, but does not interfere with the identification. Many non-metals cannot be determined by graphite furnace atomic absorption spectrometers.

The inductively coupled plasma emission spectrometer has the advantages of multi-element operation, easy coupling to the chromatograph, responsiveness to metals and many non-metals, compound-independent response, and acceptance of the continuous flow from the chromatographic column. The disadvantages are associated with the sensitivity of the plasma to organic solvents frequently used with or without admixture of water for chromatographic separations, and with the inefficiency of the nebulization and aerosol transport into the plasma. When the mobile phase becomes largely organic in nature, the plasma characteristics, and with them the emission intensities, will change causing difficulties in the quantification of analytes. The instrument should not be standardized with aqueous standard solutions, when the chromatography is carried out with an organic phase. When the mobile phase becomes rich in very volatile organic solvents the plasma may be extinguished. Some of these problems can be avoided by increasing the power at which the plasma is maintained. For example, samples of sulfur-containing surfactants could be chromatographed with methanol/water 4 : 1 (v/v) and the surfactants detected with their 180.7 nm sulfur line in an argon plasma operated at 1600 W [67]. The problems with quantification and stability of the plasma may become even more serious, when a gradient of predominantly organic mobile phases must be used for the separation. A considerable disadvantage is the high price of inductively coupled plasma emission spectrometers. However, top-of-the-line graphite furnace spectrometers do not cost much less than bottom-of-the-line sequential plasma emission spectrometers.

With respect to detection limits [50], the entire system consisting of the liquid chromatograph and the element-specific detector must be considered. The detection limits quoted for the spectrometers above will not be reached by the combined systems. During chromatography the analytes in the sample will be spread into bands from which aliquots are taken for analysis. The band-formation dilutes the analyte, and the use of aliquots from a band reduces the amount of analyte brought to the detector. For these reasons, the system detection limits will probably be ten- to fifty-times the spectrometer detection limits for a given element. When such detection limits are estimated, the fact must be remembered that spectrometer detection limits are given for analytes in benign matrices such as distilled water. Benign matrices cause few—if any—interferences. "Real" samples from the environment, for instance brines, sludges, organic-rich extracts from seaweeds, and body fluids, will contain many substances that may cause severe interferences and the real detection limits to move far away from the desirable, ideal limits. Because graphite furnace detection limits in the range of 10^{-10} to 10^{-13} and inductively coupled plasma emission spectrometer limits in the microgram per liter range are characteristic for many elements, the chromatography/element-specific detection systems achieve system detection limits that are still useful for even dirty samples.

The emphasis on the advantages and disadvantages of the element-specific detectors shall not distract from the importance of the chromatographic process in the determination and identification of trace element compounds. An analytical method will become increasingly more useful and attractive the less work on the sample prior to analysis is required. For instance, the identification of arsenic compounds in terrestrial and marine organisms would be greatly facilitated, if an aqueous or organic extract could be analyzed on a chromatograph/element-specific detector system without clean-up. Because such extracts contain many organic compounds that may interact with the analytes, their identification may become difficult, if not impossible. The retention times are the identi-

44

fying characteristics for the analytes. The retention times will change when the analytes interact with matrix components. The change in retention time with purity is evident in the chromatograms of crab extracts at two stages of clean up (Fig. 2.7). The extracts contain as the only arsenic compound arsenobetaine. Whereas the chromatogram of the highly purified extract shows the same retention time for arsenobetaine as obtained with a synthetic standard dissolved in distilled water, the impure extracts produce two peaks, one of which has a much shorter retention time [68]. This example clearly demonstrates, that

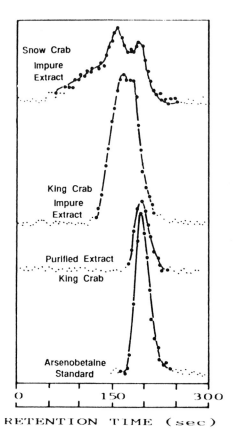

Fig. 2.7. Chromatograms of purified and impure methanolic extracts from crab meat and of a distilled water solution of synthetic arsenobetaline with an ARL 34000 simultaneous inductively coupled argon plasma emission spectrometer as the arenic-specific detector. [Hamilton PRP-1 column, Waters Associates high pressure liquid chromatograph, conditions as in ref. [11]. Redrawn from *Chemosphere* [68] by permission of Pergamon Press and the authors.

much more work is needed on the chromatographic component of the chromatograph/element-specific detection systems.

2.7. CONCLUSIONS

Several systems for the identification and quantification of trace element compounds containing metals, metalloids, and non-metals have been developed during the past two decades. These systems are based on gas chromatographic separation of volatile compounds, on the conversion of non-volatile compounds to volatile derivatives, on the liquid chromatographic separation of non-volatile compounds and the detection and quantification of these compounds by element-specific detectors. Microwave-induced plasma emission spectrometers, graphite furnace atomic absorption spectrometers, and inductively coupled argon plasma emission spectrometers serve as the most commonly used detectors in the single- or multi-element mode. Whereas better detection limits are always desirable, the available systems are well suited for the determination of trace element compounds in even "dirty" samples. Some of the efforts now directed toward the determination of total element concentrations should be dedicated to the identification of trace element compounds. Investigations of the transformations of trace element compounds in organisms and in the abiotic environment, and the study of the interactions of trace element compounds with biochemically important molecules, will lead to increased knowledge about causes of diseases, about the effects of pollutants on the environment, about better methods for treating element deficiencies and toxic episodes, and about the environmental cycles of elements. Those who use element-specific detectors cannot fail to make important discoveries and contributions.

ACKNOWLEDGEMENTS

Financial support for the work on arsenic-specific detectors and for the preparation of this publication by the Robert A. Welch Foundation of Houston, Texas is gratefully acknowledged.

REFERENCES

1   J.O. Nriagu (Ed.), Changing metal cycles and human health, Dahlem Konferenzen, Life Sciences Research Report 28, Springer-Verlag, Berlin, 1984.

46

2   W. Stumm (Ed.), Global chemical cycles and their alterations by Man, Dahlem Konferenzen, Physical and Chemical Sciences Research Report 2, Abakon Ver-lagsgesellschaft, Berlin, 1977.

3   A. Kloke, D.R. Sauerbeck and H. Vetter, in J.O. Nriagu (Ed.), Changing metal cycles and human health, Springer-Verlag, Berlin, 1984, pp. 113–141.

4   U.S. National Academy of Sciences, Series on medical and biological effects of environmental pollutants, Arsenic (1977), Chromium (1974), Copper (1977), Lead (1972), Manganese (1973, Nickel (1975), Platinum Group (1977), Selenium (1976), Vanadium (1974), Zinc (1970), Washington D.C.

5   K.J. Irgolic, J. Appl. Organomet. Chem., 2 (1988) 303–307 and references therein.

6   T.R. Irvin and K.J. Irgolic, J. Appl. Organomet Chem., 2 (1988) 509–514 and references therein.

7   M. Bernhard, F.E. Brinckman and K.J. Irgolic, in M. Bernhard, F.E. Brinckman and P.J. Sadler (Eds.), The importance of chemical "speciation" in environmental processes, Dahlem Konferenzen, Life Sciences Report 33, Springer-Verlag, Berlin, 1986, pp. 7–14.

8   Dionex IC Exchange, 7 (1988) 9.

9   D. Chakraborti, D.C.J. Hillman, K.J. Irgolic and R.A. Zingaro, J. Chromatogr., 249 (1982) 81–92.

10   K.J. Irgolic and R.A. Stockton, Mar. Chem., 22 (1978) 265–278.

11   K.J. Irgolic, R.A. Stockton, D. Chakraborti and W. Beyer, Spectrochim. Acta, 38B (1982) 437–445.

12   P.J. Craig (Ed.), Organometallic compounds in the environment: principles and reactions, John Wiley & Sons, New York, N.Y., 1986, and references therein.

13   C.J. Cappon, LC-GC, 5 (1987) 400–418.

14   B. Kolb, G. Kemmner, F.H. Schleser and E. Wiedeking, Z. Anal. Chem., 221 (1966) 166–175.

15, Y.K. Chau, P.T.S. Wong and P.D. Goulden, Anal. Chem., 47 (1975) 2279–2281.

16   D.A. Segar, Anal. Lett., 7 (1974) 89–95.

17   G.E. Parris, W.R. Blair and F.E. Brinckman, Anal. Chem., 49 (1977) 378–386.

18   F.J. Fernandez, At. Absorpt. Newsletter, 16 (1977) 33–36.

19   F.J. Fernandez, Chromatogr. Newsletter, 5 (1977) 17–21.

20   Y.K. Chau and P.T.S. Wong, Environ. Anal., (1977) 215–225.

21   J.C. van Loon, Can.J. Spectrosc., 26 (1981) 22A–32A.

22   R. Dumarey, R. Dams and P. Sandra, J. High Resol. Chromatogr., Chromatogr. Commun., (1982) 687–689.

23   A.J. McCormack, S.C. Tong and W.D. Cooke, Anal. Chem., 37 (1965) 1470–1476.

24   C.I.M. Beenakker, Spectrochim. Acta, 31 (1976) 483–486.

25   R.L. Firor, Am. Lab., May 1989, 40–48.

26   J.C. Carnahan, K.J. Mulligan and J.A. Caruso, Anal. Chim. Acta, 130 (1981) 227–241.

27   P.C. Uden, in "Developments in atomic plasma spectrochemical analysis", R.M. Barnes (Ed.), Heyden & Sons, Ltd., London, 1981, pp. 302–320.

28   P.C. Uden, Chromatogr. Forum, 1 (1986) 17–26.

29   S.W.Jordan, I.S. Krull and S.B. Smith, Jr., Anal. Lett., 15A (1982) 1131–1148.

30   I.S. Krull, S.W. Jordan, S. Kahl and S.B. Smith, Jr., J. Chromatogr. Sci., 20 (1982) 489–498.

31   R.J. Lloyd, R.M. Barnes, P.C. Uden and W.G. Elliot, Anal. Chem., 50 (1978) 2025–2029.

32   P.C. Uden, R.M. Barnes and F.P. DiSanzo, Anal. Chem., 50 (1978) 852/855.
33   D.S. Treybig and S.R. Ellebrecht, Anal. Chem., 52 (1980) 1633–1636.
34   K.W. Panaro, D. Erickson and I.S. Krull, Analyst. 112 (1987) 1097–1105.
35   D.L. Windsor and M.B. Denton, J. Chromatogr. Sci., 17 (1979) 492–496.
36   K.J. Irgolic "The determination of arsenic compounds in environmental samples" in "Hazardous metals in the environment", M. Stöppler (Ed.), Elsevier, Amsterdam, in press.
37   L. Ebdon, S. Hill, A.P. Walton and R.W. Ward, Analyst. 113 (1988) 1159–1165.
38   R. Belcher, S.L. Bogdanski, E. Henden and A. Townshend, Anal. Chem. Acta, 92 (1977) 33–41.
39   E. Henden, Anal. Chim. Acta, 173 (1985) 89–95.
40   M. Burguera and J.L. Burguera, Analyst, 111 (1986) 171–174.
41   K.W. Panaro and I.S. Krull, Anal. Lett., 17A (1984) 157–172.
42   P.J. Clark, R.A. Zingaro, K.J. Irgolic and A.N. McGinley, Intern. J. Environ. Anal. Chem., 7 (1980) 295–314.
43   D.S. Bushee, I.S. Krull, P.R. Demko and S.B. Smith, Jr., J. Liquid Chromatogr., 7 (1984) 861–876.
44   I.S. Krull and K.W. Panaro, Appl. Spectrosc., 39 (1985) 960–968.
45   T. Kaise, H. Yamauchi, T. Hirayama and S. Fukui, J. Appl. Organomet. Chem., 2 (1988) 339–347.
46   J.S. Edmonds and K.A. Francesconi, J. Appl. Organomet. Chem., 2 (1988) 297–302.
47   K.J. Irgolic, "Arsenic in the environment", in "Frontiers in bioinorganic chemistry", A.V. Xavier (Ed.), VCH Publishers, Weinheim, Germany, 1986, pp. 399–408.
48   K.J. Irgolic, R.A. Stockton and D. Chakraborti, "Determination of arsenic and arsenic compounds in water supplies", in "Arsenic: industrial, biomedical, environmental perspectives", W.H. Lederer and B.J. Fensterheim (Eds.), Van Nostrand Reinhold, New York, N.Y., 1983, pp. 282–308.
49   I.S. Krull, Trends Anal. Chem. (Pers. Ed.), 3 (1984) 76–80.
50   K.L. Jewett and F.E. Brinckman, "The use of element-specific detectors coupled with high-performance liquid chromatographs", in "Detectors in liquid chromatography". T.M. Vickrey (Ed.), Marcel Dekker Publishers, New York, N.Y., 1983, pp. 205–241.
51   K.J. Irgolic and F.E. Brinckman, "Liquid chromatography element-specific detection systems for analysis of molecular species", in "The importance of chemical speciation in environmental processes", M. Bernhard, F.E. Brinckman and P. Sadler (Eds.), Dahlem Konferenzen 1986, Springer-Verlag, Berlin, 1986, pp. 667–684.
52   C.J. Cappon, LC-GC, 6 (1988) 584–599.
53   K.J. Irgolic, Sci. Total Environ., 64 (1987) 61–73.
54   F.E. Brinckman, W.R. Blair, K.L. Jewett and W.P. Iverson, J. Chromatogr. Sci., 15 (1977) 493–503.
55   R.A. Stockton, "Graphite furnace atomic absorption and inductively coupled argon plasma emission spectrometers as element-specific detectors for the determination of trace elements and trace element compounds", Ph.D. Dissertation, Department of Chemistry, Texas A&M University, College Station, Texas, U.S.A., August 1985.
56   S.J. Haswell, R.A. Stockton, K>C.C. Bancroft, P. O'Neill, A. Rahman and K.J. Irgolic, J. Automatic Chem., 9 (1987) 6–14.
57   I.S. Krull, K.W. Panaro and L.L. Gershman, J. Chromatogr. Sci., 21 (1983) 460–472.
58   P.C. Uden and I.E. Bigley, Anal. Chim. Acta, 94 (1977) 29–34.

48

59 C.M. Kirkman, Z.B. Cheng, P.C. Uden, W.J. Stratton and D.E. Henderson, J. Chromatogr., 317 (1984) 569–578.
60 C.H. Gast, J.C. Kraak, H. Poppe and F.J.M.J. Maessen, J. Chromatogr., 185 (1979) 549–561.
61 I.S. Krull, D.S. Bushee, R.G. Schleicher and S.B. Smith, Jr., Analyst. 111 (1986) 345–349.
62 M. Morita, T. Uehiro and K. Fuwa, Anal. Chem., 52 (1980) 349–351.
63 L.T. Taylor, D.W. Hausler and A.M. Squires, Science, 213 (1981) 644–646.
64 I.S. Krull, D. Bushee, R.N. Savage, R.G. Schleicher and S.B. Smith, Jr., Anal. Lett. 15A (1982) 267–281.
65 D. Bushee, I.S. Krull, R.N. Savage and S.B. Smith, Jr., J. Liquid. Chromatogr., 5 (1982) 463–478.
66 F.E. Brinckman, K.L. Jewett, W.P. Iverson, K.J. Irgolic, K.C. Ehrhardt and R.A. Stockton, J. Chromatogr., 191 (1980) 31–46.
67 K.J. Irgolic and J.E. Hobill, Spectrochim. Acta, 42 B (1987) 269–273.
68 K.A. Francesconi, P. Micks, R.A. Stockton and K.J. Irgolic, Chemosphere, 14 (1985) 1443–1453.

I.S. Krull (Ed.), *Trace Metal Analysis and Speciation*
Journal of Chromatography Library Series, Vol. 47
© 1991 Elsevier Science Publishers B.V., Amsterdam

Chapter 3

Chromatographic Sample Introduction for Plasma Mass Spectrometry

DOUGLAS T. HEITKEMPER[1] and JOSEPH A. CARUSO[2]

[1] *Elemental Analysis Research Center, U.S. Food and Drug Administration, Cincinnati, OH 45202, U.S.A.*
[2] *Department of Chemistry, University of Cincinnati, Cincinnati, OH 45221, U.S.A.*

3.1. INTRODUCTION

Toxicologists, epidemiologists, physicians, environmental chemists and others are aware that trace elements can be harmful to the environment and many organisms at concentration levels much lower than once thought possible. In addition, they are realizing that the particular chemical species, i.e. organic, inorganic, organometallic and the oxidation state is important in determining the toxicity of an element. These observations mandate that trace elements, in their specific chemical forms, be determined reliably at picogram or even femtogram levels. The concern over these substances is well warranted since many are taken up from the environment through air, water, and soil by various plant and animal species. In addition to environmental damage and toxic consequences, there are potential carcinogenic, mutagenic, or teratogenic implications at the trace and possibly ultra-trace levels for animals and humans.

Plasma source mass spectrometry is a powerful analytical technique for trace element analysis with species selectivity when coupled with a suitable chromatographic sample introduction method. It combines the ability of the analytical plasma to atomize and ionize samples efficiently, with the sensitivity and selectivity of mass spectrometry. Following the commercial introduction of inductively coupled plasma mass spectrometry (ICP-MS) instrumentation in 1983, interest in plasma source MS increased rapidly. The enormous popularity of ICP-MS is not surprising considering the low levels of detection possible for a wide range of elements. In addition, multielement capability and the availability of isotope ratio information help make plasma source MS particularly attractive for element selective chromatographic detection.

This chapter reviews and expands on the material presented by the authors at the 1987 Eastern Analytical Symposium. A short review of the work done in this area by other researchers will be presented for completeness. The primary focus will be on chromatographic sample introduction for plasma source MS. The use of ICP-MS for the detection of high performance liquid chromatography (HPLC) eluates will be discussed. Speciation of several As and Sn compounds found in the environment demonstrates the ability of ICP-MS to provide element specific detection at ultra-trace levels. In addition, gas chromatographic detection using He microwave-induced plasma mass spectrometry (MIP-MS) is shown to be a promising technique, particularly for the determination of halogenated species.

3.2. CHROMATOGRAPHIC DETECTION USING ICP-MS

3.2.1. An overview of ICP-MS

Plasma source MS is now recognized as a very powerful technique for trace element analysis. Argon ICP has been well characterized as an emission source and has, to date, been the preferred plasma for MS. The early publications by Gray and Date [1–10], Houk, Svec and Fassel [11–16], and Douglas, French and Smith [17, 18], outlined the basic fundamentals and capabilities of this technique in which singly charged ions formed in an atmospheric-pressure plasma are extracted into a quadrupole mass analyzer for detection.

A number of excellent reviews have appeared illustrating the basic aspects of ICP-MS, as practiced today [19–24]. Recent plasma MS studies have involved basic or fundamental investigations [25–29], instrument operating modes and parameters [30–34], various applications [35–45], and chromatographic sample introduction; which will be discussed in this chapter. The primary advantages of the technique include sub-ng ml^{-1} detection limits for most elements in the periodic table. When compared with ICP-AES, ICP-MS is often 100 to 1000 times more sensitive [22]. The fast mass scanning or peak jumping routines of the quadrupole MS result in an essentially simultaneous multielement technique. In addition, isotope ratios and isotope dilution analyses are possible. Finally, the background mass spectrum is relatively simple when compared with atomic emission spectra resulting in relatively little spectral overlap.

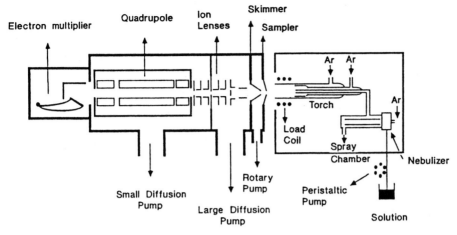

Fig. 3.1. Schematic diagram of the VG PlasmaQuad ICP-MS.

Currently, two commercial ICP-MS instruments are available, the Perkin Elmer/Sciex Elan and the VG PlasmaQuad. A schematic diagram of the PlasmaQuad instrument (the instrument in use in the authors' laboratories) is shown in Fig. 3.1. In normal operation, an aqueous sample is delivered to a pneumatic nebulizer to form an aerosol. This aerosol passes through a spray chamber to produce a more uniform droplet distribution before it reaches the plasma. Once in the plasma, desolvation, atomization, and ionization of the sample in the aerosol droplets take place. Ions are then extracted into the mass spectrometer which utilizes a three stage differentially pumped vacuum system. The plasma is sampled into the first stage through a 0.7–1.2 mm orifice centered in a nickel sampling cone. Ions enter the second vacuum stage through a skimmer cone with a similar size orifice and are directed into the quadrupole analyzer region through the use of a set of ion lenses. The quadrupole transmits ions of specific mass-to-charge (m/z) ratio, and these ions are detected using an electron multiplier. Typical operating conditions and experimental parameters are given in Table 3.1.

3.2.2. HPLC-ICP-MS

3.2.2.1. Introduction and review

Interest in element-specific detection for high performance liquid chromatography (HPLC) has increased in the past decade. A number

TABLE 3.1

TYPICAL ICP-MS OPERATING CONDITIONS

A. *Inductively coupled plasma*	
1. Incident RF power	1250 W
2. Reflected RF power	<10 W
3. Coolant argon flow rate	15 l min^{-1}
4. Auxiliary argon flow rate	<0.5 l min^{-1}
5. Nebulizer argon flow rate	0.700 l min^{-1}
6. Sampling position	On center; 10–15 mm from the load coil
B. *Vacuum system*	
1. Expansion pressure	1.5–2.5 mbar
2. Primary stage	<10^{-4}
3. Analyzer stage	5 × 10^{-6}
C. *Interface*	
1. Sampling cone	Nickel, 1 mm orifice
2. Skimmer cone	Nickel, 1 mm orifice
3. Sampler-skimmer separation	6 mm
D. *Sample introduction*	
1. Nebulizer	Concentric
2. Spray chamber	Double-pass, cooled to 5°C
3. Sample flow rate	1 ml min^{-1}

of reviews have been published describing the benefits associated with using atomic spectrometric techniques as detectors for HPLC [46–51]. Both emission [52–56] and absorption [57–60] methods have been investigated. The major requirements for coupling HPLC to any atomic spectrometric technique are compatibility of flow rate and mobile phase with the detector [48].

ICP emission techniques have been shown to be compatible with HPLC mobile phases [52–56]. The ICP readily accepts solution samples at flow rates between 0.5 and 1.5 ml min^{-1}. Additionally, many papers have been published regarding the introduction of organic solvents to the ICP [61–64]. ICP-AES also has the advantage of being a multielement technique. Thus chromatograms can be obtained while monitoring more than one element. This increases the analytical resolution available with the combined HPLC-ICP-AES technique. Additionally, chromatograms are obtained on-line in real-time as opposed to the histogram-type chromatograms typically associated with graphite furnace atomic absorption detection [47].

The major disadvantage, however, is that plasma emission detectors often lack the sensitivity for trace element levels found in many environmental and clinical samples. Often preconcentration and/or a derivitization method must be employed. The major difficulties associated with chemical speciation are the ability to collect samples representative of the system of interest and to maintain the integrity of the chemical species during sample handling and storage [50]. Therefore, it is very important to keep the amount of pretreatment to a minimum.

The greater sensitivity of plasma source MS detection often can reduce or eliminate the need for a time-consuming preconcentration and/or derivitization step [50]. Thompson and Houk [65] first reported the use of plasma MS as a detector for HPLC in 1986. They investigated the use of ICP-MS as a multielement detector for flow injection analysis and ion-pair reversed-phase HPLC. In addition to obtaining sub-ng detection limits for the speciation of As and Se compounds, they demonstrated multielement detection capabilities and the ability to measure isotope ratios on eluting peaks. This ability to measure isotope ratios with good precision (<2%) and accuracy (1%) is particularly interesting. They discussed the potential for doing stable isotope tracer studies and the advantages of quantitating species using isotope dilution. Isotope dilution would decrease analysis time compared to a calibration curve method and would help compensate for matrix effects.

Additional papers describing the use of HPLC-ICP-MS have since been published. Dean et al. [66] used size exclusion chromatography with ICP-MS detection to investigate the coupling characteristics of HPLC-ICP-MS. Using the number of theoretical plates, peak tailing, rise time, and washout time as criteria, they concluded that it was beneficial to minimize the amount of time the analyte spends as an aerosol. In addition, they demonstrated the potential usefulness of HPLC-ICP-MS for the analysis of metalloproteins.

HPLC-ICP-MS has been used for the determination of methyl mercury in an NIST reference material and thimerosal in contact lens solutions [67]. Post-column cold-vapor generation was used to increase the amount of analyte reaching the plasma and thus increased the sensitivity for Hg. Detection limits for the cold-vapor method were less than 2 ng ml^{-1} Hg.

Jiang and Houk [68] have reported on the detection of phosphorus and sulfur compounds by HPLC-ICP-MS. They used ion-pair reversed-phase chromatography for the separation of several inorganic phosphates, inorganic sulfates, nucleotides and amino acids. Organic mobile phase modifier was kept to a minimum (5%) because of the observation that analyte sensitivity decreased with increasing organic modifier concentration. The sensitivity for sulfur was low due to the use of

$^{34}S^+$ (4.2% abundant isotope). Monitoring ^{32}S (the major isotope of S) is precluded by the presence of O_2^+ at $m/z = 32$.

In a slightly different application, Jiang and co-workers [69] have used HPLC-ICP-MS for the elimination of spectral interferences caused by metal oxide formation. Molybdenum and titanium complexes of N-methylfurohydroxamic acid were retained on a polystyrene/divinylbenzene column while Cu, Zn, and Cd were eluted rapidly. This eliminated the interference of Mo and Ti oxides on Cu, Zn and Cd determinations.

There is an ever increasing interest in the effect organoarsenic, inorganic arsenic and organotin compounds have on the environment, lower animals and humans. This concern may range from how organotins affect raising and harvesting Chinook salmon in sea pens to the well being of individuals working in a smelter where inorganic arsenic is a possible hazard. The numbers of these tin and arsenic compounds which have been identified and are produced and used for many purposes is surprisingly high. Because of the toxic and environmental implications, analytical methods which allow sub-ppb detection levels (sub-ng as an absolute measure) plus a means of speciation are highly sought after and critically important to develop. HPLC-ICP-MS has been shown to provide the necessary speciation information as well as quantitation at the sub-ppb level.

3.2.2.2. *Arsenic speciation using HPLC-ICP-MS*

For many years, arsenic has been regarded as an important environmental pollutant. It can enter the environment in many ways, for example use as an agrochemical, in smelting operations, and from coal fired power plants. Buchet and Lauwerys [70] list the relative toxicities for several arsenic compounds of high concern. In descending order they give arsine > arsenite > arsenate > methanearsonic acid (MMA) > dimethylarsinic acid (cacodylic acid, DMA). Methods which can accurately assess environmental and occupational exposure to these toxic arsenic compounds are necessary.

It appears that humans deal with inorganic As by converting to the less toxic DMA or to a lesser extent MMA and then rapidly excreting it in the urine [70]. This fluid then is an important medium to monitor when considering As toxicity. Caution must be exercised, however, when using urine as a measure of exposure to inorganic As. Alessio and Bertelli [71] point out that ingestion of foods containing relatively high amounts of organoarsenicals may raise the total arsenic in urine and lead to the false assumption that it is coming from inorganic sources (the most troublesome because of the higher toxicities). Clearly when dealing with human exposure to As, total analysis is insufficient and speciation for chemical form is imperative.

To date, two publications have appeared describing the use of HPLC-ICP-MS for the speciation of arsenic in addition to the preliminary studies done by Thompson and Houk [65]. Beauchemin and co-workers [72] have used HPLC-ICP-MS to quantitate As species present in a dogfish muscle reference material. They utilized a C_{18} column with a 10 mM sodium dodecyl sulfate solution containing 5% methanol and 2.5% glacial acetic acid. The column was connected directly to the nebulizer using 0.25 mm i.d. Teflon tubing. The flow rate was 3 ml min^{-1}. A Sciex/Perkin Elmer Elan ICP-MS was used as the detector.

Under the chromatographic conditions employed, the relatively non-toxic species, arsenobetaine (the metabolic end product of arsenic for many marine animals [72]) was separated from DMA, MMA, arsenite, and arsenate. However, the two inorganic forms co-eluted along with MMA. Arsenobetaine was found to account for 84% of the total arsenic found in the dogfish muscle tissue. The absolute detection limit for arsenobetaine was reported to be 0.3 ng As.

Heitkemper and co-workers [73] have used HPLC-ICP-MS to speciate arsenic in urine. Arsenite, arsenate, DMA, and MMA were determined at the low ppb level in several urine samples. They also investigated the source of an interference believed to arise from the formation of $^{40}Ar^{35}Cl^+$ ($m/z = 75$).

The four species were separated on an amino-bonded weak anion exchange column. The mobile phase used was 30% methanol–15 mM $NH_4H_2PO_4$–1.5 mM CH_3COONH_4 and the pH was adjusted to 5.75 with glacial acetic acid. A flow rate program of 1 ml min^{-1} for 6 minutes followed by a change to 2 ml min^{-1} for the remainder of the chromatogram was used. The column was connected directly to the inlet of a concentric nebulizer with a 2-ft length of 0.010 in. i.d. Teflon tubing.

A VG PlasmaQuad ICP-MS was used with typical operating conditions and flows. Oxygen (<2%) was used in the nebulizer gas flow in order to minimize the amount of carbon build-up in the MS interface region. A Scott-type double pass spray chamber was cooled to $-5°C$ to minimize the amount of solvent vapor reaching the plasma.

Initially, a 50 mM $NH_4H_2PO_4$ mobile phase was used; however, it was found to deteriorate and clog a nickel sampling cone in approximately two hours. This prompted a reduction in phosphate concentration to 15 mM and the use of an aluminum sampler with a 0.7 mm diameter

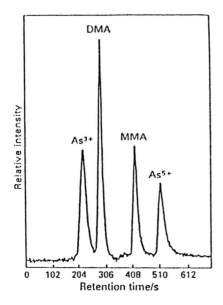

Fig. 3.2. Separation of four arsenic species by weak anion-exchange HPLC with ICP-MS detection. Each peak corresponds to 2 ng of As. Sample size, 200 μl. Mobile phase, 30% methanol–15 mM $NH_4H_2PO_4$–2 mM CH_3COONH_4. Flow-rate, 1 ml min^{-1} for 6 min followed by 2 ml min^{-1}. Monitoring, m/z = 75. Reproduced from ref. [73], *J. Anal. At. Spectrom.*, 4 (1989) 279, by permission of The Royal Society of Chemistry.

TABLE 3.2

DETECTION LIMITS AND RESPONSE DATA FOR As SPECIATION BY HPLC–ICP-MS [a]

	Arsenite	DMA	MMA	Arsenate
A. Calibration graph data [b]:				
Absolute detection limit [c] (pg As)	38	20	44	91
Sensitivity (counts pg^{-1})	5.9	11.2	5.0	2.4
Correlation coefficient	0.9999	0.9999	0.9999	0.9999
Slope, log–log data	0.972	0.975	0.988	0.992
Retention time (min)	3.7	4.7	6.9	8.5
B. Standard addition data for a representative urine sample				
Absolute detection limit (pg As)	73	38	36	96
Sensitivity (count pg^{-1})	4.1	7.0	5.9	3.3
Correlation coefficient	0.9976	0.9984	0.9982	0.9885

[a] Reproduced from ref. [73] by permission of The Royal Society of Chemistry.
[b] Four standards ranging in concentration from 10 ppb to 1 ppm; sample size = 50 μl.
[c] Detection limit = 3σ of background counts/sensitivity.

TABLE 3.3

COMPARISON BETWEEN REPORTED TOTAL ARSENIC AND DETERMINED SUM
OF ARSENIC SPECIES (μg l^{-1} $\pm$ 1 S.D.)[a]

Sample	Total As by HPLC-ICP-MS	Reported total As	Analytical method used[b]
UriChem Urine	204 $\pm$ 3[c]	220 $\pm$ 30	A
Chemistry Control		213 $\pm$ 22	B
freeze-dried urine		272 $\pm$ 36	C
		Avg.: 235 $\pm$ 29	
SRM 2670: Trace elements in freeze-dried urine; elevated level	489 $\pm$ 154	480 $\pm$ 100[d]	D, E
SRM 2670: normal level	109 $\pm$ 6	15[e]	D, E

[a] Reproduced from ref. [73] by permission of The Royal Society of Chemistry.
[b] Analytical methods: A, spectrophotometry; B, anodic stripping voltammetry; C, Zeeman-effect AAS; D, eletrothermal atomization AAS; E, instrumental neutron activation analysis.
[c] Average value for three urine samples.
[d] Certified value and 95% confidence limit.
[e] Value not certified.

orifice in place of the conventional Ni cone. The aluminum was found to tolerate the phosphate buffer better than the nickel and was used for the remainder of the work. Figure 3.2 shows a chromatogram obtained for a standard mixture of the four As species. Each of the arsenic species was present at a concentration of 10 ppb (or 2 ng absolute). The analytical figures of merit for this work are shown in Table 3.2. Detection limits ranged from 20 to 91 pg As in aqueous standards and from 36 to 96 pg As in urine. Calibration curves were linear from 10 to 1000 ppb As and the RSD's for five replicate injections were less than 6% for each of the four species.

Several urine samples were analyzed using the method of standard additions. Samples were filtered through a 0.2 micron filter before being injected onto the column. No other pretreatment or preconcentration was necessary. Two freeze-dried urine standards were used to validate the method. Agreement between reported total As concentration and the sum the concentrations of the individual species determined by HPLC-ICP-MS was within 2% and 13% for both standards spiked with arsenate. Table 3.3 shows the results of this comparison.

The determination of arsenite was complicated by the presence of an interfering peak. Investigation into the source of the interference

References pp. 70–73

showed that the peak was at least partially due to the elution of chloride from the urine and subsequent formation of $^{40}Ar^{35}Cl^+$ which interferes with the determination of As at $m/z = 75$. Chromatograms of a 1% NaCl solution were obtained at masses 35, 37, 75, and 77 in order to monitor the elution of chloride ($^{35}Cl^+$, $^{37}Cl^+$, $^{40}Ar^{35}Cl^+$, and $^{40}Ar^{37}Cl^+$). Each of the chromatograms showed peak shapes and retention times similar to the interference.

3.2.2.3. *Organotin speciation via HPLC-ICP-MS*

Organotin compounds are of high environmental and toxicological interest. They have been used extensively as biocides, catalysts and polymer stabilizers. The highly selective biocidal properties of trialkyltin compounds have resulted in many uses as fungicides, miticides, antifouling paints and wood preservatives [74]. Two prestigious groups have reviewed (1) measurements and intercomparisons [74] and (2) assessing the organotin effect of environmental quality with emphasis on the aquatic environment [75]. Both of these groups address analytical method approaches and reach the same conclusion that picomole/l (in the vicinity of tenths of ppb) analyses are necessary, plus speciation information is a requirement. For example, tributyltin ion may be toxic to some marine organisms at ppb levels while the dibutyl compound is relatively nontoxic.

Suyani and co-workers [76] have used ICP-MS for the detection of several organotin compounds separated by HPLC. Both ion exchange and ion pairing chromatography were investigated. Sub-ng limits of detection were obtained and compared with ICP-AES detection. ICP-MS was found to be three orders of magnitude more sensitive than ICP-AES.

A VG PlasmaQuad ICP-MS was used. Typical operating conditions and flows were employed throughout their work. However, the nebulizer argon flow was mixed with 3% oxygen to prevent carbon build-up on both the skimming and sampling cones resulting from the introduction of organic solvents associated with the LC mobile phase.

The interface between the chromatograph and ICP-MS consisted of a 40 cm length of 0.25 mm i.d. Teflon tubing connecting the end of the HPLC column to a Hildebrand grid nebulizer. A Scott-type double pass spray chamber was cooled to $-8°$ C in order to minimize the amount of solvent vapor reaching the plasma.

A strong cation exchange stationary phase was used with a mobile phase consisting of 0.1 M ammonium acetate in 85% v/v methanol/water.

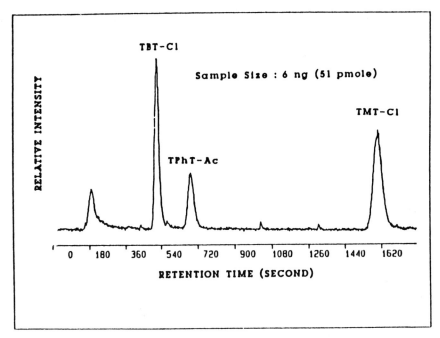

Fig. 3.3. Ion exchange chromatogram with ICP-MS detection for tin speciation. Each labeled peak corresponds to 6 ng of Sn. Strong cation exchange column. Mobile phase, 0.1 M ammonium acetate in 85% v/v methanol. Flow rate, 1 ml min^{-1}. Monitoring, m/z = 120. Reproduced from ref. [76], *J. Chromatogr. Sci.*, 27 (1989) 139, by permission of Preston Publications, a Division of Preston Industries, Inc.

The flow rate was 1 ml min^{-1}. The ion pairing work involved a C$_{18}$ column and mobile phase of 4 mM sodium pentane sulfonate in a mixture of 80 : 19 : 1 methanol/water/acetic acid. The pH was adjusted to 3.00 with the addition of 1 M sulfuric acid. A typical ion exchange chromatogram is shown in Fig. 3.3.

The figures of merit for this work are summarized in Table 3.4. Detection limits for both chromatographic modes ranged from 0.4 to 1 ng Sn. Compared with direct solution nebulization, detection limits were degraded 20 fold. Reproducibility of peak area measurements was found to be below 10% RSD. Linear dynamic ranges (LDR) were 2–3 orders of magnitude. Smaller LDR's for the ion pair chromatography were attributed to poor resolution.

Recently, Suyani and co-workers [77] have investigated the use of micellar liquid chromatography (MLC), as an alternative to to the methods given above, with ICP-MS detection. The organic solvents commonly used in HPLC mobile phases can decrease sensitivity due to excessive

TABLE 3.4

DETECTION LIMITS AND RESPONSE DATA FOR Sn SPECIATION BY HPLC–ICP-MS[a]

	TMT-Cl[c]	TPhT-Ac[d]	TBT-Cl[e]
Ion-pair chromatography			
Detection limit (ng Sn)[b]	0.4	1	0.7
Correlation coefficient	0.9999	–	0.9999
Log–log slope	0.9872	–	1.0251
Linear dynamic range	2	–	2
Reproducibility of peak area (RSD)	7.2	6.3	5.8
Ion exchange chromatography			
Detection limit (ng Sn)	0.4	1	0.8
Correlation coefficient	0.9998	0.9999	0.9992
Log–log slope	0.9955	1.0036	0.9928
Linear dynamic range	3	3	3
Reproducibility of peak area (RSD)	6.6	8.1	8.8

[a] Reproduced from ref. [76] by permission of Preston Publications, a Division of Preston Industries, Inc.
[b] Detection limit = 3σ background/slope calibration curve.
[c] TMT-Cl = trimethyltin chloride; [d] TPhT-Ac = triphenyltin acetate; [e] TBT-Cl = tributyltin chloride.

solvent loading in the plasma [61]. In MLC, organic mobile phases are replaced by aqueous solutions of micelles. Thus micellar mobile phases are an attractive alternative to conventional hydro-organic mobile phases. MLC separations have been shown to be comparable to HPLC separations using conventional hydro-organic mobile phases [78, 79].

Similar instrumentation to that described previously was used with the following changes. A switching valve was placed after the chromatographic column in order to allow the introduction of a 1% v/v nitric acid wash to the nebulizer between chromatographic runs. The mobile phase was directed to waste during this time. In addition, a Leeman-type ICP torch with demountable injector (3 mm i.d. tapered to 2 mm at the tip) was used. A nebulizer gas flow of 1.09 l min^{-1} was required in order to produce the central channel in the plasma necessary for sampling by the MS. Finally, no addition of oxygen to the nebulizer flow was necessary.

Sodium dodecyl sulfate (SDS) was used to make the micellar mobile phase. The SDS concentration was kept at or below 0.1 M in order to prevent the sampling orifice of the ICP-MS from clogging. The injector of a conventional ICP torch was found to clog after approximately 1 hour when SDS was introduced to the ICP-MS. However, similar problems were not noticed when using ICP-AES detection (in ICP-AES, the torch is mounted vertically rather than horizontally as in ICP-MS). The combination of the Leeman torch/injector and using the post-column switching valve were found to eliminate the clogging problem. No deposition occurred in the torch or sampling orifice over an 8 hour continuous run.

Figure 3.4 shows a typical MLC-ICP-MS chromatogram. A 0.1 M SDS micellar mobile phase was used with a C_{18} column for the separation of trimethyltin chloride (TMT-Cl), triethyltin bromide (TET-Br), and tripropyltin chloride (TPrT-Cl). Additionally, monomethyltin trichloride (MMT-Cl), dimethyltin dichloride (DMT-Cl), and trimethyltin chloride (TMT-Cl) were separated using a 0.02 M SDS mobile phase. The flow rate was 1 ml min^{-1} for both separations.

Figures of merit for the two separations are shown in Table 3.5.

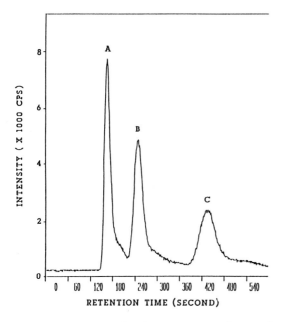

RETENTION TIME (SECOND)

Fig. 3.4. MLC-ICP-MS chromatogram of a standard mixture of trialkyltin compounds. Each peak corresponds to 4 ng Sn. A. TMT-Cl; B. TET-Br; C. TPrT-Cl. Stationary phase, C-18. Mobile phase, 0.10 M SDS, 3% v/v acetic acid, and 3% v/v propanol. Flow rate, 1 ml min^{-1}. Monitoring, $m/z = 120$. Reproduced from ref. [77], *Appl. Spectrosc.*, 43 (1989) 962–967, with permission from *Applied Spectroscopy*.

References pp. 70–73

TABLE 3.5

DETECTION LIMITS AND RESPONSE DATA FOR Sn SPECIATION BY MLC–ICP-MS[a]

	MMT-TCl[c]	DMT-DCl[d]	TMT-Cl[e]
Methyltin compounds			
Detection limit (pg Sn)[b]	46	26	126
Correlation coefficient	0.999	1.000	0.999
Log–log slope	1.00	0.95	0.96
Linear dynamic range	4	4	3.5
Reproducibility of peak height (RSD)	3.4	3.8	4.8
	TMT-Cl	TET-Br[f]	TPrT-Cl[g]
Trialkyltin separation			
Detection limit (pg Sn)	27	51	111
Correlation coefficient	1.000	0.999	1.000
Log–log slope	0.97	0.96	0.97
Linear dynamic range	4	4	3.5
Reproducibility of peak area (RSD)	1.9	1.3	1.9

[a] Reproduced from ref. [77] by permission of *Applied Spectroscopy*.
[b] Detection limit = 3σ background/slope calibration curve.
[c] MMT-Cl = monomethyltin trichloride; [d] DMT-DCl = dimethyltin dichloride; [e] TMT-Cl = trimethyltin chloride; [f] TET-Br = triethyltin bromide; [g] TPrT-Cl = tripropyltin chloride.

A comparison of detection limits was made between MLC, ion-pairing chromatography, and ion-exchange chromatography for trimethyltin chloride. The detection limit with MLC was 15 times lower than the two modes of chromatography employing hydro-organic mobile phases. At the same time, the background at $m/z = 120$, the major isotope of tin, was found to be approximately 100 counts s^{-1} compared with approximately 11,000 counts s^{-1} with ion-pair chromatography employing a 70% methanol mobile phase.

3.2.3. GC-ICP-MS

In contrast to the MIP source, ICP's have received limited attention as detectors for gas chromatography. One reason for this includes the fact that ICP's have traditionally been used for the determination of metals. The analytical utility of the ICP for the detection of non-metals such as carbon and the halogens is limited by the relatively small degree of ionization for these elements in an argon plasma. Additionally, peak broadening from the interface between capillary GC column and plasma is a concern. Relatively few reports of ICP-AES detection for GC have appeared [80–86].

Chong and Houk [87] have reported on the coupling of ICP-MS to packed-column GC. Potentially this hyphenated technique would not only provide sensitive element-selective detection, but also would allow for the determination of atomic ratios or empirical formulas. The information derived would be complementary to molecular information available using GC-MS with conventional ionization methods.

The interface between GC column and ICP consisted of a glass-lined stainless-steel tube which was attached to the end of the column. The other end of the tubing was inserted into the quartz injector tube of the ICP torch. The end of the transfer line was positioned approximately 2 cm. from the tip of the injector in order to eliminate arcing of the plasma to the stainless-steel tubing. An auxiliary argon nebulizer flow was necessary to form an ion rich central channel in the plasma.

Standard commercial mixtures of alcohols and aromatic hydrocarbons were used to evaluate the GC-ICP-MS system. Detection limits ranged from 0.001 to 400 ng s^{-1} for the elements B, C, Br, Cl, Si, P, O, I, and S. Elements with lower ionization potentials such as P and I were found to have the best detection limits. Oxygen had the poorest detection limit due to a combination of its high ionization potential and high background at $m/z = 16$. Linear dynamic ranges were 2 to 5 decades, depending primarily on background levels. Elemental ratios were investigated for C/Cl, C/Br, C/I, C/P, and C/O. The best ratios were found with C/Cl while the worst with C/O. The ratios were better when the two elements had similar sensitivities and masses. Isotope ratios were determined and the results ranged from 0.4% RSD for Br to 18% RSD for N. Chong and Houk [63] report that the isotope ratios determined on GC peaks were about a factor of three worse than the best available by conventional ICP-MS.

3.3. CHROMATOGRAPHIC DETECTION USING MIP-MS

3.3.1. An introduction to MIP-MS

For some years the potential of the MIP He source for selective chromatographic detection has been recognized. To date most studies have involved element selective detection for gas chromatography using

He MIP with optical emission detection [88–94]. Only recently has the He MIP's capability as an ion source for mass spectrometry been investigated [97].

In the early 80's, Douglas and co-workers [17, 18] described an argon MIP-MS system. Their early results were excellent. However, there are two major advantages to using a He plasma rather than Ar. Helium plasmas have a higher ionization energy than Ar (24.5 eV vs. 15.8 eV for Ar) which results in more efficient ionization of elements with higher ionization potentials such as F, Cl, Br, I, P, S and Hg [95]. In addition, the use of He should eliminate or reduce some of the spectral interferences arising from the three isotopes of argon and polyatomic species containing argon [95]. For example, the major isotopes of Ca, Fe, and Se cannot readily be determined with argon ICP-MS due to the Ar induced interferences at $m/z = 40$, 56 and 80 ($^{40}Ar^+$, $^{40}Ar^{56}O^+$, $^{40}Ar_2^+$). Montaser, Chan and Koppenaal [96] have successfully coupled an atmospheric-pressure He ICP to a commercial ICP-MS system. Although no chromatography has been reported with this system, it's potential as a chromatographic detector should not be overlooked.

3.3.2. He MIP-MS detection of gas phase halogens

Satzger and co-workers [97] used an atmospheric-pressure He MIP at moderate powers (360 W) with a laboratory constructed plasma MS instrument to detect Br, Cl and I as positive ions. In order to reduce air entrainment in the plasma plume, a torch was designed which provided a tangential sheath flow of nitrogen. The major background ions in the He mass spectrum were found to be $^{14}N^{16}O^+$ and $^{16}O_2^+$. To further reduce the entrainment of atmospheric gases, a quartz bonnet was placed between the microwave cavity and the face of the vacuum chamber. The combined use of the bonnet and sheath gas was found to effectively shield the plasma from the influx of atmospheric gases. The principal background ions with this combination, $^{14}N^+$, $^{14}N_2^+$ and $^{14}N_3^+$, were thought to originate from the N_2 sheath flow.

Increased sensitivity for Br, Cl and I as positive ions was reported when using the bonnet and sheath gas. The apparent ionization potential of the plasma seemed to have been increased above 9.25 eV (the ionization potential for NO) when using the combination [97]. The ionization potential for $^{14}N_2$ (the principal background ion in the shielded plasma) is 15.6 eV while the first ionization potentials for Br, Cl and I are 11.84, 13.01 and 10.45 eV, respectively.

Gaseous mixtures of CH_3Br, CH_3Cl and CH_3I in helium were used for this preliminary study. The detection limits reported were 1.2, 21, and 1.8 pg s^{-1}, respectively. These results were encouraging for they showed an improvement over Ar ICP-MS for Br and Cl and compare well with results obtained using He plasmas with optical emission spectrometry.

The detection of gas phase halogens by He MIP-MS has also been investigated by Brown and co-workers [98]. They utilized a commercially available ICP-MS instrument. The ICP torch box was removed and replaced with an internally tuned TM_{010} microwave cavity. The He plasma was formed using a demountable tangential flow torch. This was used in an effort to create a central region in the plasma, rich in analyte ions which allows for maximum sensitivity when sampling the plasma with the mass spectrometer.

The He background spectrum consisted of major ions at $m/z = 8$, 16, and 14 corresponding to $^4He_2^+$, $^{16}O^+$ and $^{14}N^+$, respectively. The helium background was found to be relatively free of spectral interferences above $m/z = 40$. Background species were studied with respect to their position in the plasma. Generally, intensity was maximum at small sampling depths (approximately 1 mm between the tip of the sampling cone and the faceplate of the microwave cavity). Evidence for the formation of positive ions of F, Cl, and Br was presented in the form of mass spectra showing peaks at the proper m/z values. Analyte sampling position studies showed that, generally, optimal analytical response was found at small sampling distances and centered within the plasma. Detection limits of 0.08, 0.05, and 0.14 pg s^{-1} were estimated for Cl, Br and I. No detection limit was given for F due to the use of PTFE in the construction of the MIP torch insert.

3.3.3. Gas chromatographic detection via He MIP-MS

Mohamad and co-workers [99] have used He MIP-MS for the detection of halogenated hydrocarbons separated by capillary gas chromatography. The experimental set-up used was similar to that used by Brown et al. [98] with the addition of a gas chromatograph. A schematic diagram of the GC-MIP-MS system is shown in Fig. 3.5. Several minor changes in the MIP-MS system were made. A tangential flow torch utilizing a stainless steel insert was found to provide a method for minimizing sample condensation by heating the capillary column up to the end of the insert. The use of a 0.4 mm orifice sampling cone was used to minimize the entrainment of air, which can result in the formation of polyatomic interferences. However, no comparison of background levels

66

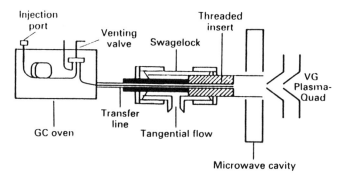

Fig. 3.5. Interface of gas chromatograph to helium MIP-MS. Reproduced from ref. [100], *J. Anal. At. Spectrom.*, 3 (1988) 923, with permission from The Royal Society of Chemistry.

with a 0.9 or 1 mm orifice was presented.

The optimum power was approximately 200 W depending on the element of interest. Sampling position was generally found to be optimal in the center of the plasma and at small sampling depths (tip of sampler even with the face plate of the cavity). Tangential He flows were kept between 6 and 8 l min^{-1}.

A transfer line interfaced the chromatograph to the MIP-MS. The capillary GC column was threaded through a 30 in. length of 1/16 in. stainless steel tubing and into the stainless steel insert of the MIP torch. The capillary column terminated flush with the end of the insert about 2 cm. from the plasma. The stainless tubing was wrapped with a high temperature heating cord. This transfer line was maintained at 250°C.

The halogenated compounds were separated using a 40 m × 0.329 mm column with an SE-54 stationary phase. Helium flow through the column was held constant at 3 ml min^{-1}. Typically, temperature programming was used with the initial temperature ranging from 70 to 110°C depending on the element being monitored. The final temperature was 250°C and the ramp was 32°C min^{-1}. Signal-to-background ratio at the major isotopes of Cl, Br, and I (*m/z* = 35, 79, 127) was used to optimize the sampling position and ion lens settings. This was accomplished using chromatographic conditions which resulted in long tailing peaks. An example GC-MIP-MS chromatogram for brominated hydrocarbons is shown in Fig. 3.6.

Analytical figures of merit for Cl, Br and I species analyzed by GC-MIP-MS are given in Table 3.6. The detection limits for chlorinated

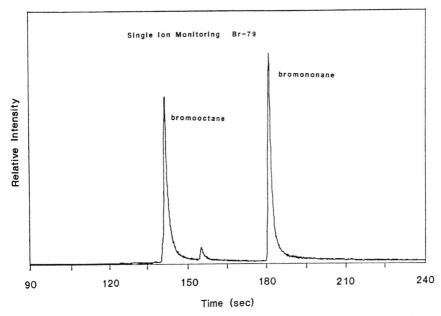

Fig. 3.6. GC-MIP-MS chromatogram of bromooctane and bromononane, 1 ng of each component injected. Stationary phase, SE-54. $T_1 = 80°C$, $T_2 = 250°C$, $Ramp = 32°C$ min^{-1}. Column flow rate, 3 ml min^{-1}. Reproduced from ref. [99], *Appl. Spectrosc.*, 43 (1989) 1127–1131, with permission from *Applied Spectroscopy*.

TABLE 3.6

DETECTION LIMITS AND RESPONSE DATA FOR HALOGENATED HYDROCARBONS BY GC-MIP-MS [a]

Compound	Linear range (decades)	Sensitivity (cps/pg)	Log–log slope	Detection limit (pg element)
Chlorobenzene	2.5	72.1	0.999	9.2
p-Chlortoluene	2.5	60.2	1.01	11.0
o-Chlorophenol	2.5	31.9	0.996	21.0
2-Bromooctane	>3	130	0.990	1.08
1-Bromononane	>3	154	0.997	0.92
Iodobenzene	>3	57.3	0.990	1.5

[a] Reproduced from ref. [99] by permission of *Applied Spectroscopy*.

species ranged from 9 (2 pg s^{-1}) to 22 pg absolute. This is at least one order of magnitude improved over MIP-AES detection in these laboratories. The relatively large background levels at masses 35 and

References pp. 70–73

37 degrade detection limits and LDR's for chlorinated species. The background at $m/z = 35$ was found to be approximately 6000–8000 counts per second (cps) while at $m/z = 79$ (^{79}Br) the background was only 200 cps. For I and Br species detection limits were found to be approximately 0.17 pg s^{-1}, at least two orders of magnitude lower than obtained in these laboratories using MIP-AES detection. LDRs ranged from 2.5 orders of magnitude for chlorinated species to 3.5 orders for some brominated and iodinated compounds.

This work demonstrates that He MIP-MS is an excellent single element detector for capillary gas chromatography. Future work should investigate the multielement capabilities of the mass spectrometer. The possibility of obtaining elemental ratios is exciting. In addition, investigation into the source and reduction of high background at low masses could result in improved detection of species containing chlorine, fluorine and phosphorus.

One approach to the problem of high background at low masses with MIP-MS is to use a reduced pressure MIP. This should completely eliminate the effect of atmospheric gases on the performance of the detector system. Creed and co-workers [100] have coupled a low-pressure MIP to the mass spectrometer.

A schematic diagram showing the low pressure design is shown in Fig. 3.7. An Ultra-Torr fitting was permanently attached to a nickel sampling cone and the plasma discharge tube was sealed from the atmosphere using this fitting. The single stage rotary pump on the mass spectrometer then evacuated the first stage along with the low pressure torch. The operating pressure was 9.0×10^{-2} mbar.

The low-pressure system was evaluated by comparing background

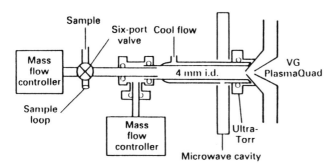

Fig. 3.7. Low-pressure He MIP-MS torch interface. Reproduced from ref. [100], *J. Anal. At. Spectrom.*, 3 (1988) 923, with permission from The Royal Society of Chemistry.

TABLE 3.7

A COMPARISON OF HELIUM GRADES AND THEIR CORRESPONDING BACK-
GROUNDS AT MASSES 35 AND 37 [a]

Gas purity (%)	Torch design	Mass 35/ counts s^{-1}	Mass 37/ counts s^{-1}	Isotope ratio
99.95	Tangential	95,000	37,000	2.56
99.95	Low pressure (trapped) [b]	4,700	3,800	1.23
99.95	Low pressure (trapped) [c]	5,700	1,800	3.18
99.99999	Low pressure (no trapping)	2,200	950	2.31

[a] Reproduced from ref. [100] by permission of The Royal Society of Chemistry.
[b] Trapping includes moisture, oxygen, and hydrocarbons.
[c] Trapping includes moisture and oxygen.

counts at m/z = 35, 37, 31, and 56 (Cl, P, and Fe) with those obtained using the atmospheric-pressure system. Table 3.7 shows the results obtained for masses 35 and 37. Several steps were taken to purify the helium plasma gas. Moisture, oxygen, and hydrocarbon traps were used. In addition, research-grade He (99.99999%) was used without any trapping. The low-pressure design resulted in greatly reduced background counts; however, the research-grade He showed a further decrease in background counts indicating that the elevated background at low mass is at least partially due to impurities in the He gas. The use of liquid helium boil-off may be a relatively inexpensive solution to the problem since research-grade He is expensive to use on a regular basis. Preliminary results for the detection of methyl chloride showed improved sensitivity for Cl over the atmospheric-pressure system.

3.4. CONCLUSIONS AND FUTURE DEVELOPMENTS

The use of plasma mass spectrometry for the detection of chromatographic eluates has been discussed. The combination of HPLC with ICP-MS detection provides an analytical technique capable of providing species selectivity and sub-ng levels of detection. The superior detection levels, as compared to atomic emission techniques, should spur the development of both HPLC and GC procedures capable of speciating trace elements in environmental and clinical samples. Furthermore, the use of a helium microwave plasma source has been described. Preliminary studies have demonstrated the potential advantages of He MIP-MS for GC detection of halogenated compounds.

References pp. 70–73

Chromatographic detection using plasma mass spectrometry exhibits tremendous potential; however, further studies are necessary. Multielement detection and isotope ratio determinations on eluting peaks are of considerable interest. In addition, the use of plasma sources other than the argon ICP should prove beneficial for the determination of non-metal species separated by HPLC. Finally, both sample introduction to the plasma and the interface between the chromatograph (both HPLC and GC) and plasma mass spectrometer should be areas of future research.

REFERENCES

1 A.L. Gray, Anal. Chem., 47 (1975) 600.
2 A.L. Gray, Dyn. Mass Spectrom., 4 (1976) 153.
3 A.L. Gray, Dyn. Mass Spectrom., 5 (1979) 106.
4 A.R. Date and A.L. Gray, Analyst (London), 106 (1981) 1255.
5 A.L. Gray and A.R. Date, Dyn. Mass Spectrom., 6 (1981) 252.
6 A.L. Gray and A.R. Date, Int. J. Mass Spectrom. Ion Phys., 46 (1983) 7.
7 A.R. Date and A.L. Gray, Spectrochim. Acta, Part B, 38B (1983) 29.
8 A.R. Date and A.L. Gray, Analyst (London), 108 (1983) 159.
9 A.R. Date and A.L. Gray, Int. J. Mass Spectrom. Ion Phys., 48 (1983) 357.
10 A.L. Gray and A.R. Date, Analyst (London), 108 (1983) 1033.
11 R.S. Houk, V.A. Fassel, G.D. Flesch, H.J. Svec, A.L. Gray and C.E. Taylor, Anal. Chem., 52 (1980) 2283.
12 R.S. Houk, H.J. Svec and V.A. Fassel, Appl. Spectros., 35 (1981) 380.
13 R.S. Houk, V.A. Fassel and H.J. Svec, Dyn. Mass Spectrom., 6 (1981) 234.
14 R.S. Houk, V.A. Fassel and H.J. Svec, Org. Mass Spectrom., 17 (1982) 240.
15 R.S. Houk and J.J. Thompson, Am. Mineral., 67 (1982) 238.
16 R.S. Houk, A. Montaser and V.A. Fassel, Appl. Spectros., 37 (1983) 425
17 D.J. Douglas and J.B. French, Anal. Chem., 53 (1981) 37.
18 D.J. Douglas, E.S.K. Quan and R.G. Smith, Spectrochim. Acta, Part B, 38B (1983) 39.
19 R.S. Houk, Anal. Chem., 58 (1986) 97A–105A.
20 R.S. Houk and J.J. Thompson, Mass Spectrom. Rev., 7 (1988) 425–461.
21 A.L. Gray, in F. Adams, R. Gijbels and R. VanGrieken (Editors), Inorganic Mass Spectrometry, Wiley, New York, 1988, Ch. 6.
22 G. Horlick, S.H. Tan, M.A. Vaughan and Y. Shao, in A. Montaser and D.W. Golightly, Inductively Coupled Plasmas in Analytical Atomic Spectrometry, VCH Publishers, New York, 1987, pp. 361–398.
23 A.L. Gray, Spectrochim. Acta, Part B, 40B (1985) 1525–1537.
24 D.J. Douglas and R.S. Houk, Prog. Anal. At. Spectrosc., 8 (1985) 1–18.
25 M.A. Vaughan and G. Horlick, Appl. Spectros., 40 (1986) 434–445.
26 S.H. Tan and G. Horlick, Appl. Spectros., 40 (1986) 445–460.
27 S.H. Tan and G. Horlick, J. Anal. At. Spectrom., 2 (1987) 745–763.
28 J.A. Olivares and R.S. Houk, Anal. Chem., 57 (1985) 2674– 2679.
29 J.A. Olivares and R.S. Houk, Anal. Chem., 58 (1986) 20–25.

30 S.E. Long and R.M. Brown, Analyst (London), 111 (1986) 901–906.

31 G. Zhu and R.F. Browner, Appl. Spectros., 41 (1987) 349–359.

32 G. Horlick, S.H. Tan, M.A. Vaughan and C.A. Rose, Spectrochim. Acta, Part B, 40B (1985) 1555–1572.

33 G.P. Russ III and J.M. Bazan, Spectrochim. Acta, Part B, 42B (1987) 49–62.

34 R.S. Houk and J.J. Thompson, Biomed. Mass Spectrom., 10 (1983) 107.

35 D. Beauchemin, J.W. McLaren, A.P. Mykytiuk and S.S. Berman, Anal. Chem., 59 (1987) 778.

36 J.W. McLaren, A.P. Mykytiuk, S.N. Willie and S.S. Berman, Anal. Chem., 57 (1985) 2907–2911.

37 A.R. Date and A.L. Gray, Spectrochim. Acta, Part B, 40B (1985) 115–122.

38 B.T.G. Ting and M. Janghorbani, Anal. Chem., 58 (1986) 1334.

39 B.T.G. Ting and M. Janghorbani, Spectrochim. Acta, Part B, 42B, (1987) 21.

40 H.P. Longerich, B.J. Fryer and D.F. Strong, Spectrochim. Acta, Part B, 42B (1987) 39.

41 H.P. Longerich, B.J. Fryer and D.F. Strong, Spectrochim. Acta, Part B, 42B (1987) 101.

42 S.J. Jiang and R.S. Houk, Spectrochim. Acta, Part B, 42B (1987) 93

43 J.R. Garbarino and H.E. Taylor, Anal. Chem., 59 (1987) 1568.

44 J.W. McLaren, D. Beauchemin and S.S Berman, Anal. Chem., 59 (1987) 610–613.

45 T.D.B. Lyon, G.S. Fell, R.C. Hutton and A.N. Eaton, J. Anal. At. Spectrom., 3 (1988) 265–271.

46 J.C. Van Loon, Anal. Chem., 51 (1979) 1139A.

47 I.S. Krull, Trends in Analytical Chemistry, 3 (1984) 76.

48 K.J. Irgolic and F.E Brinckman, Dahlou Conference Report, Berlin, Sept. 2–7, 1984, Springer Verlag, 1986.

49 P.C. Uden, Trends in Analytical Chemistry, 6 (1987) 238.

50 P.E. Gardiner, J. Anal. At. Spectrom., 3 (1988) 163.

51 C.J. Cappon, LC/GC, 6 (1988) 584.

52 J.P. McCarthy, J.A. Caruso and F.L. Fricke, J. Chromatogr. Sci., 21 (1983) 389.

53 K.J. Irgolic, R.A. Stockton, D. Chakraborti and W. Beyer, Spectrochim. Acta, Part B, 38B (1983) 437.

54 D.S. Bushee, I.S. Krull, P.R. Demko and S.B. Smith Jr., J. Liq. Chromatogr., 7 (1984) 861.

55 C.H. Gast, J.C. Kraak, H. Poppe and F.J.M.J. Maessen, J. Chromatogr. 185 (1979) 549.

56 I.S. Krull and K.W. Panaro, Appl. Spectrosc., 39 (1985) 960.

57 F.E. Brinckman, K.L. Jewett, W.P. Iverson, K.J. Irgolic, K.C. Ehrhardt and R.A. Stockton, J. Chromatogr., 191 (1980) 31.

58 R.A. Stockton and K.J. Irgolic, Int. J. Environ. Anal. Chem., 6 (1979) 313.

59 L. Ebdon, S.J. Hill and P. Jones, Analyst, 110 (1985) 515.

60 K.L. Jewett, F.E. Brinckman, J. Chromatogr. Sci., 21 (1981) 583–593.

61 A.W.Boorn and R.F. Browner, Anal. Chem., 54 (1982) 1402–1410.

62 P. Barrett and E. Pruszkowska, Anal. Chem., 56 (1984) 1927–1930.

63 M.W. Blades and B.L. Caughlin, Spectrochim. Acta, Part B, 40B (1985) 579–591.

64 T.J. Brotherton, J.T. Creed, D. Heitkemper, P.E. Pfannerstill, S.E. Pratsinis and J.A. Caruso, J. Anal. At. Spectrom., 4 (1989) 341–345.

65 J.J. Thompson and R.S. Houk, Anal. Chem., 58 (1986) 2541–548.

66 J.R. Dean, S. Munro, L. Ebdon, H.M. Crews and R.C. Massey, J. Anal. At. Spec-

72

trom., 2 (1987) 607–610.

67 D.S. Bushee, Analyst, 113 (1988) 1167–1170.

68 S.J. Jiang and R.S. Houk, Spectrochim. Acta, Part B, 43B (1988) 405–411.

69 S.J. Jiang, M.D. Palmieri, J.S. Fritz and R.S. Houk, Anal. Chim. Acta, 200 (1987) 559–571.

70 J.P. Buchet and R. Lauwerys, in S. Facchetti (Editor), Analytical Techniques for Heavy Metals in Biological Fluids, Elsevier, Amsterdam, 1981, pp. 75–90.

71 L. Alessio and G. Bertelli, in S. Facchetti (Editor), Analytical Techniques for Heavy Metals in Biological Fluids, Elsevier, Amsterdam, 1981, pp. 41–64.

72 D. Beauchemin, M.E. Bednas, S.S. Berman, J.W. McLaren, K.W.M. Siu and R.E. Sturgeon, Anal. Chem., 60 (1988) 2209–2212.

73 D. Heitkemper, J. Creed, J. Caruso and F.L. Fricke, J. Anal. At. Spectrom., 4 (1989) 279–284.

74 W. Blair, G. Olsen, F. Brinckman, R. Paule and D. Becker, "An International Butyltin Measurement Methods Intercomparison:..", NBSIR 86-3321, Nat'l. Bureau of Standards, Dept. of Commerce, February 1986.

75 J. Thompson, M. Sheffer, R. Pierce, Y. Chau, J. Cooney, W. Cullen and R. Mcquire, "Organotin Compounds in the Aquatic Environment:...", NRCC #22494, National Research Council, Canada, Assoc. Comm. Environ. Safety, 1985.

76 H. Suyani, J. Creed, T. Davidson and J. Caruso, J. Chromatogr. Sci., 27 (1989) 139–143.

77 H. Suyani, D. Heitkemper, J. Creed, and J. Caruso, Appl. Spectros., 43 (1989) 962–967.

78 J.G. Dorsey, M.T. DeEchegaray and J.S. Landy, Anal. Chem., 55 (1983) 924.

79 J.S. Landy and J.G. Dorsey, Anal. Chim. Acta, 178 (1985) 179.

80 L. Ebdon, S. Hill and R.W. Ward, Analyst, 111 (1986) 1113.

81 D.L. Windsor and M.B. Denton, Appl. Spectrosc., 32 (1978) 366.

82 D.L. Windsor and M.B. Denton, Anal. Chem., 51 (1979) 1116.

83 D.L. Windsor and M.B. Denton, J. Chromatogr. Sci., 17 (1979) 492.

84 R.M. Brown, Jr. and R.C. Fry, Anal. Chem., 53 (1981) 532.

85 R.M. Brown, Jr., S.J. Northaway and R.C. Fry, Anal. Chem., 53 (1981) 934.

86 J.M. Keane, D.C. Brown and R.C. Fry, Anal. Chem. 57 (1985) 2526.

87 N.S. Chong and R.S. Houk, Appl. Spectros., 41 (1987) 66–74.

88 M.L. Bruce and J.A. Caruso, Appl. Spectrosc., 39 (1985) 942.

89 D.L. Haas and J.A. Caruso, Anal. Chem., 57 (1985) 846.

90 A.H. Mohamad and J.A. Caruso, in J.C. Giddings (editor), Advances in Chromatography, Marcel Dekker, New York, 1987, pp. 191–227.

91 L. Ebdon, S. Hill and R.W. Ward, Analyst, 111 (1986) 1113.

92 D.R. Luffer, L.J. Galente, P.A. David, M. Novotny and G.M. Hieftje, Anal. Chem., 60 (1988) 1365–1369.

93 S.R. Goode, B. Chambers and N.P. Buddin, Appl. Spectrosc., 37 (1983) 439.

94 J.C. Evans, K.B. Olsen and D.S. Sklarew, Anal. Chim. Acta, 194 (1987) 247–260.

95 J.T. Creed, T.M. Davidson, W.L. Shen, P.G. Brown and J.A. Caruso, Spectrochim. Acta, Part B, 44B (1989) 909–924.

96 A. Montaser, S.K. Chan and D.W. Koppenaal, Anal. Chem., 59 (1987) 1240–1242.

97 R.D. Satzger, F.L. Fricke, P.G. Brown and J. Caruso, Spectrochim. Acta, Part B, 42B (1987) 705–712.

98 P.G. Brown, T. M. Davidson and J. Caruso, J. Anal. At. Spectrom., 3 (1988) 763–769.

99   A.H. Mohamad, J.T. Creed, T.M. Davidson and J.A. Caruso, Appl. Spectros., 43 (1989) 1127–1131.
100   J.T. Creed, A.H. Mohamad, T.D. Davidson, G. Ataman and J.A. Caruso, J. Anal. At. Spectrom., 3 (1988) 923–926.

I.S. Krull (Ed.), *Trace Metal Analysis and Speciation*
Journal of Chromatography Library Series, Vol. 47
© 1991 Elsevier Science Publishers B.V., Amsterdam

Chapter 4

The Future of Intelligent Spectrometers in Speciation by Atomic Emission Spectrometry

ROBERT B. BILHORN [1], ROBERT S. POMEROY [2], and M. BONNER DENTON [2]

[1] *Eastman Kodak Company, Kodak Park Building #34, Rochester, NY 14650, U.S.A.*
[2] *University of Arizona, Department of Chemistry, Tucson, AZ 85721, U.S.A.*

4.1. INTRODUCTION

Analytical chemists are interested in qualitative analysis, quantitative analysis, and speciation. Today's analytical chemist would like to have an instrument into which any type of sample can be introduced and the instrument would provide all the information requested about that sample. Such an analyzer should give the requested answer with absolute certainty and completely free from any type of matrix effects. Ideally, it would even return the sample unharmed. In fact, the ideal analytical technique should be economical, fast, reliable, simple to operate, and require very small samples, and possibly be nondestructive. Plasma spectroscopic techniques, while far from being such a conceptually ideal analyzer, can contribute to our knowledge of elemental constituents and their quantization, but by themselves, these techniques do not provide insight into speciation. Chromatography provides a separation in time of closely related chemical species, but many of the detection schemes used with chromatography offer limited information. The power of combining chromatographic separation techniques with element specific detection can help provide great insights into speciation. Chromatographic methods have been used as introduction techniques into plasmas, particularly microwave and inductively coupled plasmas for some years [1–22]. Figure 4.1 shows a system configured for element selective determination of the effluent from a gas chromatograph [12].

Figures 4.2 and 4.3 depict the simultaneous output of the individual wavelength detectors monitoring the emission of carbon, hydrogen, and silicon lines [23], and the carbon, hydrogen, and tin lines [12]. Each set

References pp. 98–99

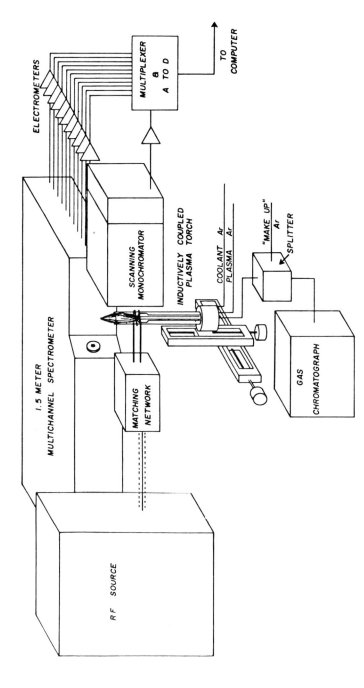

Fig. 4.1. Experimental system for simultaneously monitoring the multiple wavelengths associated with transitions of a compounds elemental components.

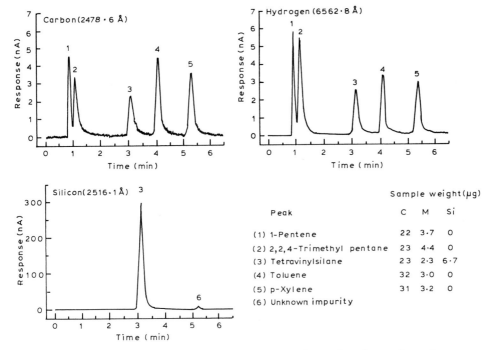

Fig. 4.2. Simultaneous analog plots of the emission intensity observed at the wavelengths of carbon, hydrogen and silicon.

of these chromatograms is obtained from one injection, i.e., the carbon trace, the hydrogen trace, and the tin trace were all obtained from the same chromatographic separation. Digitizing a series of points over each one of the peaks and then calculating the relative ratios for each of the different components at each sequential concentration across the peak, first as the peak raises to the highest concentration and then decreases past the peak maximum, allows accurate empirical formula determination. Table 4.1 provides the relative empirical formulas for a series of hydrocarbons, the theoretical ratios, the observed ratios, and absolute and relative errors [24]. While this configuration was capable of measuring a large number of metals, carbon, hydrogen, and certain other nonmetals, it was not capable of determining all other nonmetals. Since the resonant transitions of a number of nonmetals, principally nitrogen, oxygen, chlorine, bromine, and sulfur, provide emission lines in the vacuum ultraviolet, this particular system was not capable of observing those resonant transitions. A series of studies was initiated to ascertain the feasibility of observing the resonant transitions

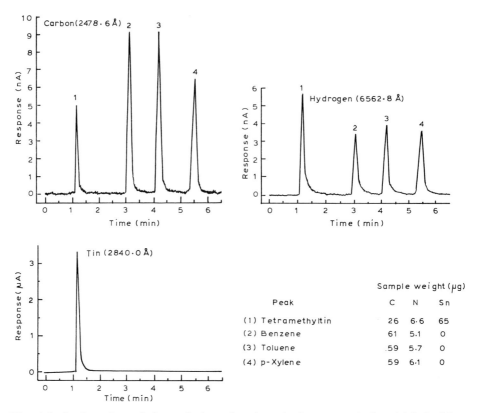

Fig. 4.3. Analog plots of the emission of carbon, hydrogen, and tin which had been rapidly digitized and reconverted into analog form.

TABLE 4.1

EMPIRICAL FORMULAS FOR A VARIETY OF HYDROCARBONS DEMONSTRATING ACCURATE DETERMINATION OF THE HYDROGEN TO CARBON RATIOS

	Empirical formula	H/C atomic ratio		Difference (%)	
		theoretical	found	absolute	relative
Cumene	C_3H_4	1.333	1.376	0.043	3.23
Cyclohexene	C_3H_5	1.667	1.646	0.021	1.26
Ethylbenzene	C_4H_5	1.250	1.249	0.001	0.08
n-heptane	C_7H_{16}	2.286	2.299	0.013	0.57
Isooctane	C_4H_9	2.250	2.260	0.010	0.44
Methylcyclohexane	CH_2	2.000	1.999	0.001	0.05
1-pentene	CH_2	2.000	1.957	0.043	2.15
o-xylene	C_4H_5	1.250	1.249	0.001	0.08
m-xylene	C_4H_5	1.250	1.263	0.013	1.04

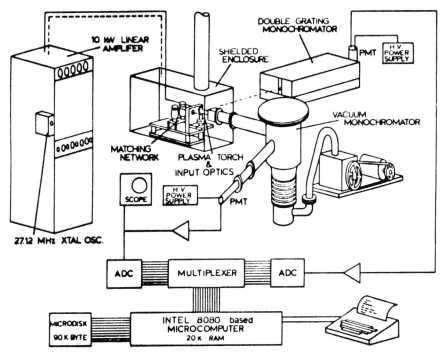

Fig. 4.4. Experimental configuration used for initial vacuum ultraviolet investigations. Later studies substituted a vacuum polychromator for the scanning vacuum monochromator.

of these nonmetals in the vacuum ultraviolet. Figure 4.4 shows the configuration used for initial studies in the vacuum ultraviolet region [25]. This system was later modified by replacing the monochromator with a vacuum polychromator. The modification involved installing a demountable inductively coupled plasma torch that has been fitted with a side arm which is purged with helium or argon, a magnesium fluoride lens which transfers visible, ultraviolet, and vacuum ultraviolet light into either the monochromator or polychromator. Working curves for oxygen at 130 nm, nitrogen at 149 nm, bromine at 153 nm, and chlorine at 134 nm are shown in Fig. 4.5 [26].

Figure 4.6 demonstrates the separation of a multicomponent mixture showing the determination of bromine, chlorine, nitrogen, oxygen and carbon.

Other studies that had been ongoing for some period of time contributed another key aspect important for using the inductively coupled plasma as an element selective detector for liquid chromatography.

References pp. 98–99

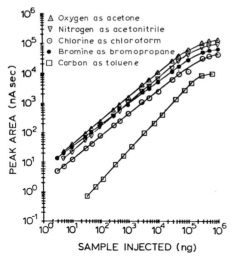

Fig. 4.5. Working curves obtained for oxygen ($\triangle$) at 130.49 nm, nitrogen ($\triangledown$) at 149.28 nm, chlorine (o) at 134.72 nm, bromine ($\bullet$) at 153.17 nm, and carbon ($\square$) at 247.8 nm.

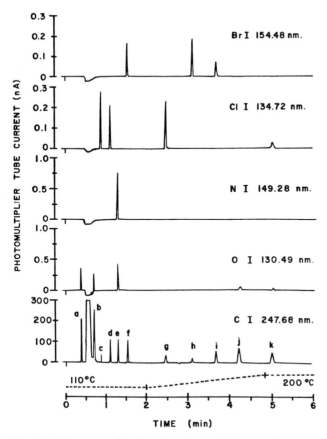

Fig. 4.6. Element selective detection of the gas chromatographic separation of a mixture of ethanol (a), i-propanol (b), carbon tetrachloride (c), 1,2-dichloroethane (d), nitroethane (e), n-bromobutane (f), tetrachloroethylene (g), 1,2-dibromoethane (h), bromobenzene (i), phenol (j), and chlorobenzaldehyde (k).

These investigations had been aimed at developing nebulizers capable of handling difficult matrix solutions, including samples containing high levels of suspended materials and saturated salt solutions. These nebulizers are also capable of stable operation for long periods of time, aspirating more moderate salt levels, such as buffer solutions [27–29]. The studies of the original Babington geometry nebulizer had evolved into a variety of different types of nebulizers, including miniature Babington nebulizers, various types of channels, and pool types of nebulizers. These investigations have demonstrated the ability to analyze rather difficult types of matrices varying from and including motor oils, condensed milk, whole blood, hydrolytic fluid, urine, orange juice, pineapple syrup, tomato sauce, and saturated salt solutions. Using the nebulizer shown in Fig. 4.7, the configuration shown in Fig. 4.8 was implemented [30].

This is a conventional high performance liquid chromatograph with the output of the UV detector connected into the nebulizer for the inductively coupled plasma. Figure 4.9 compares the output of the UV detector with that of the ICP which, in this case, shows only phosphorus-containing compounds. Figure 4.10 shows the determination of adenosine triphosphate (ATP), adenosine diphosphate (ADP), uridine monophosphate (UMP), guanosine monophosphate (GMP) and cytidine monophosphate (CMP) by observing phosphorus emission at

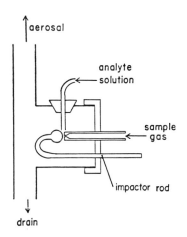

Fig. 4.7. Miniature Babington principle nebulizer used to aerosolize effluent from an HPLC. While inefficient at generating usable aerosol, this geometry proved to be stable over long periods of time, even when used with relatively high salt concentration mobile phases.

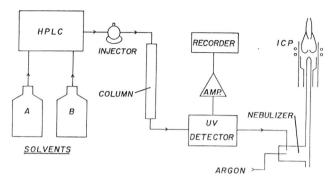

Fig. 4.8. Diagram of the HPLC-ICP system used in the described investigations.

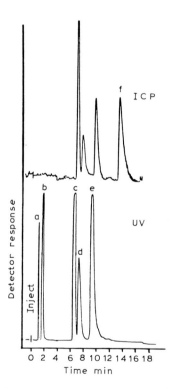

Fig. 4.9. Comparison of the signals obtained from the UV detector and phosphorus emission signal for the separation of three nucleotides using a 5-min linear gradient solution: solvent peak (a), nucleoside (b), AMP (c), CMP (d), UMP (e), and $H_2PO_4^-$ (f). Peak profiles remained relatively unchanged between the UV and emission systems indicating that, under these conditions, the dead volume associated with the nebulizer was insignificant.

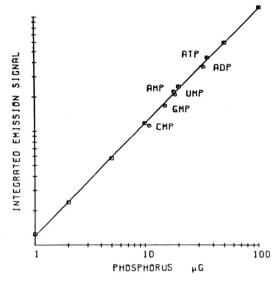

Fig. 4.10. Results of nucleotide analysis plotted on a phosphorus calibration curve.

213.6 nm [30]. Quantitation is achieved by either standardization with any one of the nucleotides or by merely injecting a known amount of any other phosphorus containing compound. Inexpensive, stable primary standards containing an element can be utilized to develop the working curve, i.e., sodium phosphate, and then quantitate unavailable compounds. Clearly, the preceding discussion indicates the great potential of combining chromatographic separation with true multiwavelength detection from the vacuum ultraviolet to visible and beyond. One additional very important aspect of plasma emission spectroscopy, widely known by all workers but seldom employed, is the fact that most elements have more than one emission line. Many have a very large number of emission lines. The conventional slew-scan instrument, while capable of measuring any combination of these lines, is really not suitable for use with chromatographic separation when one needs to monitor more than one element and, therefore, more than one wavelength simultaneously. A more viable alternative has been the direct reader or polychromator where a series of slits, photomultiplier detectors, and readout amplifiers can simultaneously monitor the emission signals occurring at several different wavelengths. In practice, generally it is impractical to dedicate more than one photomultiplier tube for one element. Therefore, one is limited to observing only a single wavelength for a particular element. Problems with the polychromator are widely recognized. Such systems are rather expensive, generally are fairly large and bulky, require a significant amount of initial set up time, cannot readily be realigned to different wavelengths, require somewhat specialized laboratory space, and of course, are able to acquire only a very limited amount of spectral data, since one slit and photomultiplier tube detector are required for each wavelength observed.

4.2. MULTICHANNEL ARRAY DETECTORS

Over the years, people have dreamed about replacing the direct reader with some type of electronic readout, which can measure the photon flux at all wavelengths simultaneously. The electronic equivalent of the photographic emulsion. A variety of camera devices have been explored for this application, including vidicons, intensified target vidicons, plumbicons, orthicons, image dissectors, photodiode arrays, and a number of other types of imaging devices. These camera techniques virtually always suffered from one or more of the following problems [31–34].

(1) Small dynamic range — In atomic spectroscopy, concentrations are commonly quantitated between 4 and 6 orders of magnitude. Additionally, there are emissions which can exceed the analyte either from the plasma background or major matrix components.

(2) Insufficient spectral range — Most workers would be happy with 185 nm to maybe 500 or 600 nm. However, ideally one would like to cover from the vacuum ultraviolet region of approximately 125 nm to 1 μm or even beyond.

(3) Poor reproducibility between detector elements.

(4) Crosstalk between the elements — This includes both blooming and smearing problems.

(5) Insufficient number of resolution elements

(6) Poor signal-to-noise ratios

(7) Inability to integrate photon flux

(8) Inability to randomly access detector elements

(9) High cost for detector elements

(10) Poor reliability

4.3. CHARGE TRANSFER DETECTORS

For some years, astronomers have been developing a new class of light detection technology, charge transfer devices. These devices are solid state, multichannel, integrating photon detectors which either employ inter- or intra-charge transfer readout. Two of the most common types of these devices are the charge coupled device and the charge injection device [35–38]. Properly operated, these charge transfer devices can provide extremely high quantum efficiency over extended wavelength ranges, extremely low dark count rates, low read noises, a high degree of durability, and today are available in a large variety of formats. Properly operated scientific charge transfer devices can outperform photomultiplier tubes in terms of signal-to-noise performance, dynamic range, cost, and of course, they are multichannel devices. Today's best photomultiplier tubes exhibit peak quantum efficiency near 30 percent. In contrast, many charge transfer devices have peak quantum efficiencies exceeding 50 percent, and some devices are able to achieve almost 90 percent quantum efficiency over limited wavelength regions. The combined characteristics of low readout noise, low dark current, and high quantum efficiency allow charge transfer detectors to outperform photomultiplier tubes, in cases where the integration time approaches that normally used for atomic spectroscopic analysis.

Today, the highest performance charge transfer devices are the charge coupled device technologies. However, these devices have two significant problems when employed in atomic emission spectroscopy. The charge coupled device is inherently not a random accessible readout system. In order to interrogate a specific wavelength region, the entire device must be scanned. Additionally, many of today's charge coupled devices have problems associated with blooming. This phenomenon results when a high photon flux falls on a small area of the detector, resulting in a very large number of charge carriers. Once a certain level of charge carriers is reached, the charge carriers spill into adjacent detector sites. This spillage can continue into more and more detector sites causing large areas of device to become unusable as a result of only one very intense feature [39].

In contrast, the charge injection device does not suffer from these blooming problems [40, 41]. Additionally, it is either random or pseudo-randomly addressable and each detector site can be read in either a nondestructive or destructive fashion. This extraordinarily powerful capability allows the host computer to interrogate a given detector site, determine if a significant signal has accumulated at that particular location to provide the desired analytical information, or if additional integration will be required. Hence, with the charge injection device, it is possible to continuously interrogate the entire spectral region being observed, while the analysis is actually proceeding to determine the

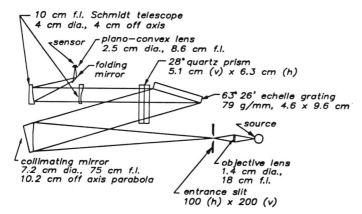

Fig. 4.11. Optical diagram of the echelle spectrometer using a custom designed off axis Schmidt image reduction configuration to provide an optical focal plane suitable for use with a CID array detector.

optimal integration time for each observed wavelength. Unfortunately, the charge transfer devices are not readily available in geometries suitable for use in conventional optical spectrometers. In fact, while charge transfer devices are available in the hundreds to millions of detector elements, most larger devices are rectangular arrays and each detector element is relatively small. Conventional linear spectrographic dispersion systems are not suitable. The echelle grating spectrograph first developed by Harrison [42] can provide relatively high resolution and the desired rectangular format. Current commercially available echelle systems, however, do not generate a sufficiently small spectral display for use with reasonably sized charge transfer devices. Therefore, to most effectively utilize the charge transfer device detectors, custom optical systems are necessary. A number of optical configurations have been investigated [43–47]. Figure 4.11 shows a successful optical geometry developed at the University of Arizona capable of supplying the appropriate imaging format and the spot quality necessary to effectively utilize the charge transfer device detection technologies [48].

This system uses an off-axis parabolic mirror, an echelle grating, quartz prism combination along with a specially designed off-axis Schmidt corrected image reducing telescope. The off-axis feature minimizes vignetting that was observed with our previous Schmidt telescope designs. Figure 4.12 shows the system block diagram which utilizes a CID Tech CID17BAS (Liverpool, NY) operated in a specially designed camera system and readout electronics. The readout electronics are con-

CID17/Echelle System Block Diagram

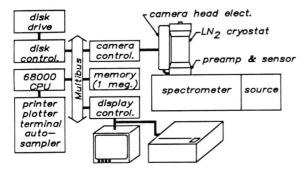

Fig. 4.12. Diagram of the echelle AES system used in conjunction with the optics of Fig. 4.11. The detector array is cooled to below 90 K to virtually eliminate dark current.

trolled by a modified Photometrics (Tucson, AZ) Model 2700026 camera controller which is hosted in a multibus based Motorola 68000 system.

Figure 4.13 shows an iron spectra taken with the system without background correction. Figure 4.14 shows the same iron spectra where a blank has been run previously and subtracted out. Notice that a very large amount of band structure and plasma emission have been eliminated. The overall design criteria for this system required that 80 percent of the light from a single emission line fall on an array of 3×3 detector elements. This criteria has easily been achieved, and in fact, in some cases, over 80 percent of the light can actually be observed on a single pixel. Individual orders are horizontally oriented

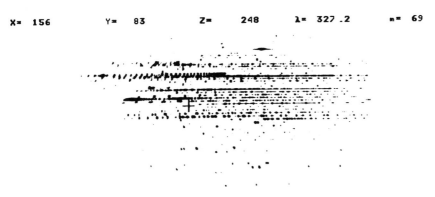

Fig. 4.13. An echellogram of iron before the plasma background is subtracted. Long wavelength orders are at the top with wavelength decreasing on each succeeding order. Note the cursor is on the 69th order at 327.2 nm.

References pp. 98–99

Fig. 4.14. The echellogram of Fig. 4.13 following removal of the plasma background.

and orders are stacked vertically (Figs. 4.13 and 4.14). Low orders, i.e., long wavelengths are at the top of the spectrum with the ultraviolet at the center and bottom of the spectrum. Figure 4.15A shows a plot of a small area of a CID device where a portion of three orders are visible [34]. Emission features are apparent in each one of the stripes. However, determination of which of these features are background or analyte is not easy. Figure 4.15B shows the background for the exact same region. The background is shown subtracted in Fig. 4.15C. The large peak was clearly due to analyte, however, there are several other features clearly above the signal/background ratio which can be used for qualitative and quantitative determinations.

As previously mentioned, the charge injection device has a very unique feature in its ability to read out a given detector element nondestructively. This capability can be used to achieve a number of important goals. Charge transfer devices possess a source of noise not encountered with the conventional photomultiplier tube. This noise is known as read noise. Merely reading the device adds additional noise to the signal. In the case of the charge coupled device (CCD), this noise has been pushed to extraordinary low levels. Today, scientifically operated CCDs are capable of providing read noise levels of less than 5 electrons. Current charge injection devices have read noises that are much higher. In fact, their single read noise is on the order of 1,000 electrons. While this is considerably less than most photodiode arrays, it is still a very significant problem at very low light levels. However, the nondestructive readout mode can readily be employed to significantly decrease this read noise component. Figure 4.16 shows the

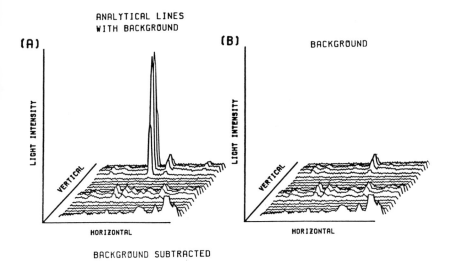

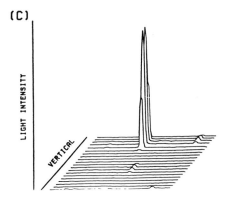

Fig. 4.15. (A) Plot of a small section of an echellogram showing analytical lines with background from the plasma; (B) plot of the background from the plasma; and (C) plot of the background subtracted from the analytical signal. Several low intensity lines are easily observed.

result of summing multiple rereads. In this case, the read noise, which is basically a white noise source, can be reduced by the square root of the total number of reads. Summing nondestructive reads can therefore achieve read noise levels on the order of 50 electrons. Immediately the question arises of how nondestructive is the nondestructive read mode. Figure 4.17 shows the results of observing a plasma for 30 seconds, shutting the entrance slit, and reading four different intensity lines for 8 hours nondestructively [49]. Note there is no perceivable decay in the signal and also note there is no perceivable dark current.

References pp. 98–99

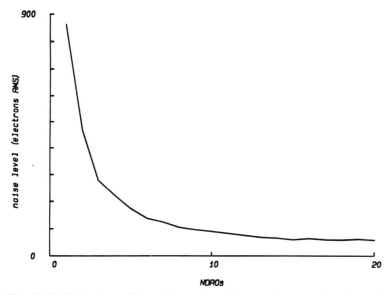

Fig. 4.16. Reduction of the observed readout noise as a function of summing multiple nondestructive readouts. The noise of 860 electrons was reduced to 60 electrons following averaging 400 rereads.

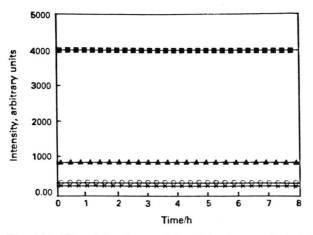

Fig. 4.17. Plot of the observed signal for four analytical lines of varying intensity read nondestructively at a rate of one read per second for 8 hours. Note that the observed intensities are not affected by over 2.8×10^5 readouts.

Detection limits achieved for a number of elements are compared with those from the ARL Beckman DC plasma echelle system in Table 4.2 [50]. Clearly, the detection limits are very comparable. The detection

TABLE 4.2

DETECTION LIMITS OBSERVED WITH THE DESCRIBED CID ECHELLE SPEC-
TROMETER

Element	Wavelength (nm)	Detection limit (ppb)	
		UACID17 Echelle	Beckman spectra Span III
Al [a]	308.22	7	80
	394.40	3	
			2 (396.15)
Ba	413.07	4	
	455.40	0.5	0.3
	393.37	1	0.7
Ca [b]	396.85	2	
	422.67	0.4	
	357.87	2	2
	359.35	4	
Cr [ab]	360.53	6	
	427.48	4	
	428.97	5	
Cs	852.12	200	
	894.36	10	
			800 (455.53)
Cu	324.75	3	2
	327.40	3	2
	338.50	80	
	345.40	16	
Dy [a]	353.20	6	
	353.60	10	
	364.54	2	
	421.20	3	
	358.12	6	
	371.99	1	5
Fe	373.49	3	
	373.71	15	
	386.00	15	
	287.42	50	
Ga [c]	294.36	5	
	403.30	4	
	417.21	5	

TABLE 4.2 (continued)

Element	Wavelength (nm)	Detection limit (ppb)	
		UACID17 Echelle	Beckman spectra Span III
	339.90	40	
	341.65	30	
	345.60	10	
	345.70	10	
Ho [a]	347.43	20	
	348.48	7	
	379.68	4	
	389.10	3	
	405.39	1	
	410.38	5	
	303.94	7	20
In	325.86	12	
	410.18	2	10
	451.13	2	4
	279.55	0.2	0.2
Mg	280.27	0.4	0.6
	285.21	0.6	
	259.37	5	2
	293.93	2	
Mn [ab]	403.08	4	10
	403.31	5	
	403.45	6	
Ni [b]	352.45	7	
			2 (341.48)
Pb [b]	363.96	11	
	405.78	6	20
			10 (368.35)
	331.20	30	
Th	332.50	30	
	346.90	70	
	374.10	10	
			20 (401.91)
	294.19	800	
U	367.01	40	
	385.96	300	100
	393.20	25	
			40 (424.17)

TABLE 4.2 (continued)

Element	Wavelength (nm)	Detection limit (ppb)	
		UACID17 Echelle	Beckman spectra Span III
	409.90	90	
	412.80	100	
V	437.92	40	5
	438.47	3	
	439.00	15	
			2 (309.31)
	289.10	25	
Yb [a]	369.40	2	
	398.80	2	

[a] Detection limit determined as part of the three element mix Al, Cr and Mn.
[b] Detection limit as part of the five element mix Ca, Cr, Mn, Ni and Pb.
[c] Beckman detection limit data not available for this element.

limit data for a number of elements tabulated for the CID system were run in mixtures, while the published commercial detection limits, no doubt, were achieved through optimization of individual elements. The ability to observe, nondestructively, the emission occurring at each different analyte wavelength during the very early stages (milliseconds) of each analysis provides the ability to make a variety of intelligent decisions while the actual analysis is proceeding. Wavelengths which are most appropriate for a particular element can be determined based on the observed concentrations for that element at that particular time in the analysis, the observed concentrations for other interfering analytes, the host matrix solvent system, etc. A variety of readout modes are available and the optimal readout mode can be determined based on how many detector elements are to be read for each wavelength for the signal and background correction, and what data reduction modes will be employed. Diagnostic procedures can also be employed. The argon emission from the plasma can be utilized to determine if the sample is perturbing the excitation conditions. Additionally, either the hydroxyl band emission or hydrogen atomic emission can be utilized to monitor nebulizer performance and analyte temperature. During the analysis, the system can choose the optimal integration time for each analytical wavelength, can collect appropriate background data for each analytical wavelength for the exact same period that one is to observe

94

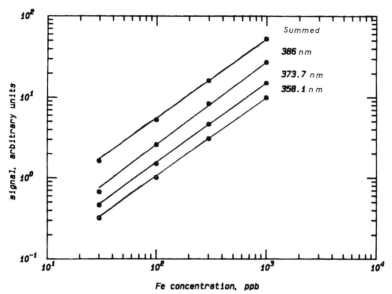

Fig. 4.18. Observed working curves for the 386 nm, 373.7 nm and 358.1 nm iron lines and the resultant curves when the three lines are summed.

TABLE 4.3

EFFECTS OF SUMMING SIGNALS FROM MULTIPLE EMISSION WAVELENGTHS FROM AN ELEMENT ON THE LIMIT OF DETECTION AND SENSITIVITY

Element and wavelengths	Number of of lines summed	LOD for summed lines $(ng\ ml^{-1})$	Increase in LOD (%)	Increase in sensitivity (%)
Mg 279.6, 280.3, 285.2 nm	3	0.1	60	110
Fe 358.1, 373.7, 386.0 nm	3	7	50	90
Sr 421.6, 460.7 nm	2	8	30	50
Dy 364.5, 353.2 nm	2	8	30	95
Ho 345.6, 379.7, 389.1 nm	3	2	35	70
Yb 369.4, 398.8 nm	2	0.3	25	50

that analytical wavelength, and can monitor all the various diagnostic parameters. Following the analysis, this system can reduce and present the data, and can compare the results obtained for each element at each wavelength employed for that element. (Note that whenever possible, multiple emission wavelengths are simultaneously observed for each element.) The resulting data base can be used to estimate the accuracy and precision for a given analysis. Any unusual circumstances that were encountered during an analysis can be noted and appropriate follow-up procedures can then be recommended. The concept of using more than one wavelength to increase sensitivity and detection limits is shown in Fig. 4.18, where three relatively equivalent iron emission lines are summed to improve system performance. Table 4.3 shows the concept of summing lines on the increase in limit of detection and in sensitivity.

4.4. ARRAY DETECTION OF CHROMATOGRAPHIC ELUENTS

Application of CID based detection to simultaneous multielement emission analysis of chromatographic eluents, involves the measurement of the emission intensity of spectral lines corresponding to each of the elements to be determined over fixed time intervals. The charge integration time used is set at a constant value chosen, such that the shape of eluting peaks is not significantly degraded. The array is cleared of all charge between integrations, and integrations are performed one immediately after another. The data are presented as the integrated signal, as a function of the number of short time intervals, resulting in a chromatogram with the appearance of one recorded with a conventional detector. Increased sensitivity is achieved by sacrificing peak shape and increasing the time interval.

A second mode of operation offers greater sensitivity and involves integrating the entire chromatogram. In this mode, nondestructive readouts are used at regularly spaced time intervals to sample the signal charge, but the charge is allowed to continue integrating throughout the separation. The output appears as the integral of a normal chromatogram. Care must be exercised when employing this mode of operation, however, in order to avoid saturating the detector before the end of the separation.

In the typical chromatographic experiment, the operator selects the time interval between points and the total length of time to be allowed for acquiring the chromatogram. An analysis specification file is then created by selecting the elements to be determined and the spectral lines

References pp. 98–99

to be used. Once this is done, an injection can be made. The spectrometer waits for the operator to strike a key indicating that a sample has been injected onto the column. The first few seconds after the key is struck are used for measuring the spectral background at each of the selected spectral lines. The integration of signals from spectral lines is then immediately started. If a spectral line reaches the threshold level before the end of the integration period, then it is recorded and its exact integration time stored. All other spectral lines are recorded at the end of the integration period and their exact integration times stored. Once all spectral lines have been recorded, the signal levels are normalized to the same, pre-selected, integration interval; the background signal recorded at the beginning of the chromatogram is subtracted, and the spectral line area is integrated. The CID is then cleared of charge and the process repeated until the entire chromatogram has been recorded.

Verification of the methodology was achieved by using the echelle spectrometer system as an element selective detector for both anion exchange and reverse phase HPLC. The anion exchange work involved the separation and detection of the ethylenediaminetetraacetic acid (EDTA) and nitrilotriacetic acid (NTA) complexes of Ca, Cu, and Mg, duplicating the work of Fraley et al. [51]. The reverse phase HPLC work consisted of the separation of acetylacetonate (AcAc) complexes of a number of transition metals employing a C-18 column and a 50 : 50 methanol water mobile phase.

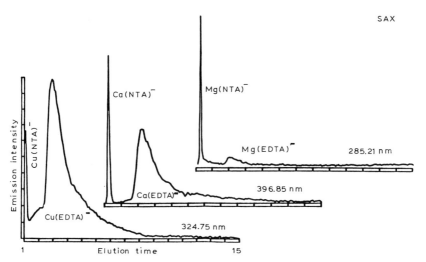

Fig. 4.19. Simultaneous atomic emission detection using the CID array spectrometer of the complexes of Ca, Cu, and Mg following anion exchange chromatography.

Figure 4.19 shows the chromatograms recorded when a 20 μl sample containing 10 ppm each of the species $Ca(NTA)^-$, $Ca(EDTA)^{2-}$, $Cu(NTA)^-$, $Cu(EDTA)^{2-}$, $Mg(NTA)^-$, and $Mg(EDTA)^{2-}$ was injected onto a 2 cm dry packed SAX column. In this chromatogram, 5-second integrations were employed and the original peak shapes were reproduced relatively unaltered, as verified by the output of a UV absorption detector connected in series with the spectrometer.

The sensitivity of the AES detection method is on the order of tens of nanograms, as determined by measurement of the signal-to-noise ratio observed in the chromatograms in Fig. 4.19. This is in agreement with the detection limits measured in solution by normal atomic emission. The use of chromatographic separation prior to AES detection can now provide insight into the attached ligand and/or oxidation state of the metal. The two methodologies combined make a powerful tool in the determination of chemical speciation.

4.5. CONCLUSIONS

Future research in the area of atomic emission spectroscopic detection of chromatographic eluates employing CID technology should benefit from the tremendous flexibility in wavelength selection afforded by a CID/AES system, as well as the sensitivity offered by the CID. The CID is easily capable of following transient signals which require recording on sub-second time intervals, allowing the system to be applicable in all areas of chromatography. Element specific detection employing a plasma emission source offers high sensitivity regardless of the chemical origin of the detected elements and can greatly reduce the restrictions on the mobile phases which may be employed. Probably the single largest drawback toward obtaining the detection limits for some analysis is the poor efficiency of the nebulization systems for the plasma sources. Detection limits are degraded by the inefficiency of typical nebulizers. Current research into alternate nebulizers, such as the ultrasonic nebulizer indicate that increases in detection by a factor of 10 to 20 are possible [29].

In summation, the technology of using charge transfer device detectors for simultaneously monitoring a large number of lines in a multielement analysis can achieve higher reliability, higher sensitivity, can yield a larger linear dynamic range, and has the potential of truly being developed into an expert type of system.

References pp. 98–99

98

REFERENCES

1 C.A. Bache and D.J. Lisk, Anal. Chem. 39, 786 (1967).
2 R.M. Dagnall, S.J. Pratt, T.S. West, and D.R. Deans, Talanta 16, 797 (1969).
3 R.M. Dagnall, T.S. West, and P. Whitehead, Anal. Chim. Acta 60, 25 (1972).
4 C.A. Bache and D.J. Lisk, Anal. Chem. 37, 1477 (1965).
5 W. Braun, N.C. Peterson, A.M. Base, and M.J. Korylo, J. Chromatogr. 55, 237 (1971).
6 R.M. Dagnall, T.S. West, and P. Whitehead, Anal. Chem. 44, 2074 (1972).
7 W.R. McLean, D.L. Stanton, and G.E. Penketh, Analyst 98, 432 (1973).
8 C.I.M. Beenakker, Spectrochim. Acta B 32, 173 (1977).
9 D.L. Windsor and M.B. Denton, Appl. Spectrosc. 32, 366 (1978).
10 B.D. Quimby, P.C. Uden, Anal. Chem. 50, 2112 (1978).
11 G.F. Kirkbright, A.F. Ward, and T.S. West, Anal. Chim. Acta 62, 241 (1972).
12 D.L. Windsor and M.B. Denton, J. Chromatogr. Sci. 17, 492 (1979).
13 C.H. Gast, J.C. Kraak, H. Poppe, and F.J.M. Maessen, J. Chromatogr. Sci. 185, 549 (1979).
14 D.M. Fraley and S.E. Manahan, Anal. Chem. 51, 2225 (1979).
15 M. Morita, T. Uehiro, and K. Fuwa, Anal. Chem. 52, 349 (1980).
16 P.C. Uden, Chromatography Forum, 17–26, Nov.–Dec. (1986).
17 L. Ebdon, S. Hill, and W. Ward, Analyst 111, 1113–1138 (October 1986).
18 K.W. Panaro, D. Erickson, and I.S. Krull, Analyst 112(8), 1097 (1987).
19 K.G. Michlewicz and J.W. Carnahan, Anal. Lett. 20(8), 1193 (1987).
20 P.E. Gardiner, P. Braetter, B. Gercken, and A. Tomiak, J. Anal. Atomic Spectrom. 2(4), 375 (1987).
21 G.K.C. Low, E.G. Batley, S.J. Buchanan, Anal. Chim. Acta 197, 327 (1987).
22 P.C. Uden, Trends Anal. Chem. 6(9), 238 (1987).
23 D.L. Windsor, Ph.D. Dissertation, University of Arizona, Tucson, Arizona (1977).
24 D.L. Windsor and M.B. Denton, Anal. Chem. 51, 1116 (1979).
25 D.R. Heine, J.S. Babis, and M.B. Denton, Appl. Spec., 34, 595 (1980).
26 M.B. Denton, M.J. Pilon, and J.S. Babis, Appl. Spec., 44(6), 975 (1990).
27 R.C. Fry and M.B. Denton, Anal. Chem. 49, 1413 (1977).
28 R.C. Fry and M.B. Denton, Appl. Spec. 33 393 (1979).
29 M.B. Denton, J.M. Freelin, and T.R. Smith, "Ultrasonic, Babington and Thermospray Nebulization," in Sample Introduction in Atomic Spectroscopy, Chapter 4, Elsevier Science Publishers (1988).
30 D.R. Heine, M.B. Denton and T.D. Schlabach, Anal. Chem. 54, 81 (1982).
31 Y. Talmi, Anal. Chem. 47, 658A (1975).
32 Y. Talmi, Anal. Chem. 47, 699A (1975).
33 Y. Talmi, "Multichannel Image Detectors," ACS Symposium Series, Vol. 102, Washington, D.C. (1979).
34 Y. Talmi, "Multichannel Image Detectors, Volume 2," ACS Symposium Series, Vol. 236, Washington, D.C. (1983).
35 R.B. Bilhorn, J.V. Sweedler, P.M. Epperson, and M.B. Denton, Appl. Spec. 41(7), 1114 (1987).
36 R.B. Bilhorn, P.M. Epperson, J.V. Sweedler, and M.B. Denton, Appl. Spec. 41(7), 1125 (1987).
37 P.M. Epperson, J.V. Sweedler, R.B. Bilhorn, G.R. Sims, and M.B. Denton, Anal. Chem. 60(4), 327A (1988).

38 J.V. Sweedler, R.B. Bilhorn, P.M. Epperson, G.R. Sims, and M.B. Denton, Anal. Chem. 60(5), 282A (1988).

39 J.V. Sweedler, R.D. Jalkian, R.S. Pomeroy, and M.B. Denton, Spectrochim. Acta 44B(7), 683 (1989).

40 G.R. Sims and M.B. Denton, J. Opt. Eng. 26(10), 999 (October 1987).

41 G.R. Sims and M.B. Denton, J. Opt. Eng. 26(10), 1008 (October 1987).

42 G.R. Harrison, J. Opt. Amer. 39, 552 (1949).

43 H.L. Felkel, Jr., H.L. Pardue, Anal. Chem. 49, 1112 (1977).

44 W. Liller, Appl. Opt. 9, 2332 (1970).

45 P.N. Keliher, C.C. Wohlers, Anal. Chem. 48, 333A (1976).

46 B.C. Boland, B.B. Jones, S.F.T. Engstrom, Solar Physics 17, 333 (1971).

47 W.G. Elliott, Amer. Lab. 2, 67 (1970).

48 T.R. Smith and M.B. Denton, Appl. Spec., 43(8), 1385 (1989).

49 M.B. Denton, The Analyst 112, 347 (1987).

50 R.B. Bilhorn, Ph.D. Dissertation, University of Arizona, Tucson, Arizona (1987).

51 D.M. Fraley, D.A. Yates, S.E. Manahan, P. Stalling, and J. Petty, Appl. Spec. 35, 525 (1981).

I.S. Krull (Ed.), *Trace Metal Analysis and Speciation*
Journal of Chromatography Library Series, Vol. 47
© 1991 Elsevier Science Publishers B.V., Amsterdam

Chapter 5

Inductively Coupled Plasma-Mass Spectrometry for Element-Selective Detection in Liquid Chromatography

R.S. HOUK[1] and S.-J. JIANG[2]

[1] *Ames Laboratory, U.S. Department of Energy, and Chemistry Department, Iowa State University, Ames, IA 50011, U.S.A.*
[2] *Chemistry Department, National Sun Yat-Sen University, Kaohsiung Taiwan 80424, Republic of China*

5.1. INTRODUCTION

Since its introduction in 1980 [1], inductively coupled plasma-mass spectrometry (ICP-MS) has become a widely-used analytical technique. The major reasons for this rapid growth of ICP-MS are its excellent detection limits (1–20 ng l^{-1} for most elements), the uncomplicated mass spectra observed, and the capability for rapid measurement of isotope ratios. The major applications of ICP-MS have involved elemental analysis. The analytical capabilities of ICP-MS for elemental analysis, particularly the excellent detection limits, make it attractive as an element-selective detector for chromatography. Combining a chromatographic separation with ICP-MS could provide a technique capable of distinguishing between different chemical forms of elements, i.e., "chemical speciation." In principle, ICP-MS detection for chromatography should also permit isotope ratio measurements for individual elemental species for isotope tracer studies. Finally, chromatographic removal of interfering matrix elements would extend the range of application of ICP-MS and simplify or hasten procedures for sample preparation.

In this chapter, recent studies that illustrate these capabilities are described. The use of ICP-MS with liquid chromatography is emphasized. Some work on ICP-MS with gas chromatography and solvent extraction is also reported.

5.2. INSTRUMENTATION

5.2.1. ICP-MS device

Developments in ICP-MS are reviewed biennially in the journal *Analytical Chemistry* [2]. Detailed descriptions of ICPs [3, 4], ICP-MS devices [5–7], and applications [8] are available in the literature, so only a brief survey of the instrumentation is given here.

The ICP and ion sampling interface are depicted in Fig. 5.1. The ICP is an atmospheric pressure discharge in a flowing gas stream (usually argon). Electrical power is added inductively to the plasma from an RF current applied to the load coil. The temperatures inside the plasma vary with spatial position and are thought to differ for different species but are in the range 4000–8000 K. For ICP-MS, the plasma is generally operated horizontally. As injected sample constituents travel through the axial channel of the ICP, they are vaporized, atomized and ionized.

Since the ICP operates at atmospheric pressure, the ions must be extracted into a vacuum system before they can be mass analyzed. The interface for this extraction is shown on the right of Fig. 5.1. The ions are entrained in the flow of neutral gas through the first orifice, usually

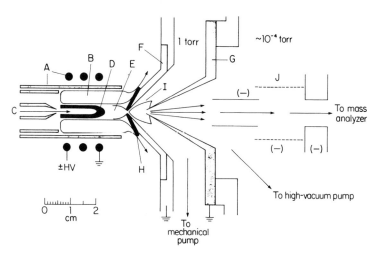

Fig. 5.1. ICP and ion sampling interface. A = torch and load coil (HV = high voltage), B = induction region of ICP, C = a solution aerosol being injected into axial channel, D = initial radiation zone, E = normal analytical zone, F = nickel cone with sampling orifice in tip, G = skimmer cone, H = boundary layer of ICP gas deflected outside sampling orifice, I = expanding jet of C gas sampled from ICP, and J = ion lens elements. Reproduced from *Anal. Chem.* [5] with permission of American Chemical Society.

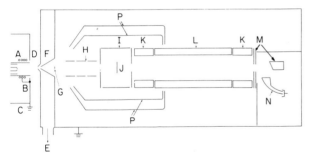

Fig. 5.2. SCIEX ELAN ICP-MS device. A = ICP, B = load coil grounded at center, C = shielding box, D = sampler, E = mechanical vacuum pump, F = skimmer, G = grounded metal stop, H = ion lenses, I = Bessel box, J = photon stop, K = RF only quadrupole rods, L = quadrupole mass analyzer, M = ion lenses and deflector, N = Channeltron electron multiplier, P = cryoshells. Reproduced from *Mass Spectrom. Reviews* with permission of John Wiley.

called the sampler. A supersonic jet forms inside the first vacuum chamber. The central section of the supersonic jet flows through the second orifice, called the skimmer. The ions that exit the skimmer are collimated and transmitted by an ion lens into a mass spectrometer (Fig. 5.2). Note that no additional ionization source is present in the interface; the ions observed are those that were in the ICP and survived the extraction process.

Most of the axial channel flows through the sampler. Roughly 1% of this sampled gas also flows through the skimmer [9, 10]. The sampling orifice is approximately 1 mm in diameter and is large enough to resist plugging by condensed solids from the sample, although such plugging can occur if the level of dissolved solute is too high. For example, solutes such as Al, Ca and Zr, which form refractory oxides, cause substantial plugging when present at 1% in the sample [11]. Plugged samplers can often be cleaned in an ultrasonic bath and re-used. Despite their exposure to the hot plasma, both sampler and skimmer are fairly rugged. They will generally survive at least 2 months of routine use before they expand or become pitted and require replacement. The commercial systems are configured with isolation valves so that the sampler and skimmer can be replaced in a few minutes.

With most present ICP-MS instruments, a quadrupole mass analyzer is used (Fig. 5.2). This device provides essentially unit mass resolution and can be scanned and switched rapidly between peaks of interest. The mass-selected ions are detected with an electron multiplier operated in either a pulse counting mode (for trace constituents) or in an analog mode for the large ion currents due to major or minor constituents. Some recent work has shown the feasibility of using a double focusing MS with an ICP for resolving atomic analyte ions from polyatomic interfering ions at the same nominal mass [12].

The mass spectrometer is also fairly robust. The electron multiplier generally needs replacement about once a year. There is little need to clean the vacuum chamber and mass filter unless they should become coated with pump oil from gross misuse. The ion lenses may require occasional cleaning to remove deposited material if highly concentrated solutions are analyzed for long periods of time. Thus, extended operation of an ICP-MS device requires more maintenance than an ICP emission spectrometer but is still not excessive.

5.2.2 The ICP as an ion source

The ICP has many desirable features as an ion source for elemental analysis. The ionization efficiency approaches 100% for elements with first ionization energies below about 8 eV, which includes most metals. Naturally, nonmetals are not as efficiently ionized, but useful numbers of positive ions can be observed even from As (ionization energy = 9.8 eV), Se (9.75 eV), and Br (11.8 eV). Multiply charged ions are not very

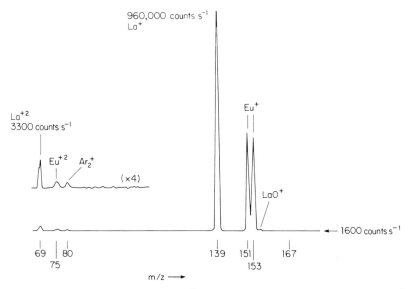

Fig. 5.3. Mass spectrum of 1 mg l^{-1} La and equimolar Eu showing M^+, M^{2+}, and MO^+. Reproduced from *Anal. Chem.* [5] with permission of American Chemical Society.

abundant, so the ICP is a very efficient but not overly energetic ion source.

These characteristics are illustrated in the mass spectrum shown in Fig. 5.3. Lanthanum at 1 mg l^{-1} yields nearly 10^6 counts s^{-1}, which is near the upper end of the linear range of the detector. Peaks are observed for La^{2+} and LaO$^+$, but they are small relative to La$^+$. The low abundance of LaO$^+$ shows that even refractory rare earth oxides are dissociated efficiently (but not H completely). Most elements yield much lower levels of M^{2+} or MO$^+$ than shown for La in Fig. 5.3. The two isotope peaks for Eu$^+$ illustrate the isotopic capability of the technique. Finally, the total signal for the two isotopes of Eu$^+$ is comparable to the signal for La$^+$ because these two elements are ionized to nearly the same extent in the ICP and because they are extracted and detected with similar efficiency.

5.2.3. Introduction of LC effluents

Liquid samples are generally nebulized and the resulting aerosols are injected into the axial channel of the ICP (Fig. 5.1). A pneumatic nebulizer, such as the concentric type shown in Fig. 5.4, is commonly used. The nebulizer is usually mounted inside a spray chamber, which prevents large droplets from reaching the plasma. In ICP-MS, the outer wall of the spray chamber is often cooled to remove as much solvent as possible [14, 15]. Removal of solvent is particularly desirable when an organic modifier is used for the separation, as described below.

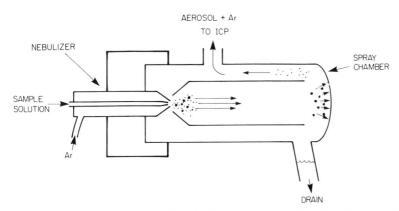

Fig. 5.4. Concentric pneumatic nebulizer and spray chamber for introducing solution into an ICP. Reproduced from *Handbook of Physics and Chemistry of Rare Earths* [13] with permission of North-Holland Physics Publishing.

Roughly 1% to 3% of the sample flow into the nebulizer actually reaches the plasma. Fortunately, the ICP-MS device has the powers of detection to compensate for this inefficiency. The nebulizer is content with sample flow rates in the range 0.3–3 ml min^{-1}. If the spray chamber outlet is connected directly to the base of the torch, as is common, the ion signal from a fresh sample can be observed roughly 3 s after the sample reaches the tip of the nebulizer.

For LC experiments, the column outlet can simply be connected to the nebulizer, and this is the approach taken by most workers. Some band broadening doubtless occurs between the column and the plasma. The general experience in the authors' laboratory is that such broadening is caused primarily in the liquid transfer line between the column and the nebulizer rather than in the gas flow through the nebulizer to the plasma. Thus, the liquid transfer line between the column and nebulizer should be narrow and short. Other workers using a UV absorbance detector followed by ICP-MS can not put the column close to the nebulizer and face a compromise between the band broadening in the liquid phase and that in the aerosol after the nebulizer. They generally prefer to minimize the length of time the sample spends as aerosol at the expense of a longer time as liquid [17–19].

The LC-ICP-MS experiments in the authors' laboratory have employed an ultrasonic nebulizer, which is roughly 10% efficient [20, 21] but has a substantial dead volume in the gas phase. When the technique progresses to the use of better separations of more complex mixtures, sample introduction devices with less opportunity for broadening, such as the direct injection nebulizer [22, 23], may become necessary.

5.2.4. Optimization of operating conditions

The sensitivity (i.e., the analyte ion count rate per unit concentration) and the background of an ICP-MS device can be quite sensitive to operating conditions such as aerosol gas flow rate, power applied to the plasma, and sampling position [24–26]. The background generally increases (particularly at m/z values below 80) and the analyte sensitivity tends to decrease as the organic content of the mobile phase increases. The presence of a metal matrix ion in the mobile phase, such as an ion-pairing reagent in reversed phase LC or an alkali buffer salt in ion exchange, can also suppress the analyte ion signal [18, 27–31]. In our experience, it is generally desirable to minimize the concentrations of both organic modifiers and matrix salts in the eluent. Mason et al. [18] report a similar observation from their experiments with size exclusion chromatography. Thus, selection of operating conditions that yield both optimum chromatographic separations and best ICP-MS performance is not always easy, and some empirical compromises between the quality of the separation and the performance of the detector may be necessary.

The dependence of both analyte signal and background on mobile phase composition complicates use of solvent gradients with ICP-MS. Thus, most separations to date have simply been isocratic. Elder and co-workers have described some work with gradient elution [19]. They have also added an additional solvent gradient post-column. The composition of this second gradient opposes that of the eluent through the column. In this way, gradient elution is used for the separation while a solvent of constant composition is nebulized into the ICP [31]. Post-column addition of an internal standard element could also facilitate use of gradient elution. Additional practical problems include significant levels of metal ions in the solvents and reagents (observable because of the high sensitivity of ICP-MS) and slow bleeding of metal species from previous samples off the column.

5.3. ELEMENTAL SPECIATION BY LC-ICP-MS

5.3.1. Separation of test mixtures

Selected ion chromatograms for $^{75}As^+$ and $^{78}Se^+$ in a mixture of species are shown in Fig. 5.5. This separation is performed by anion

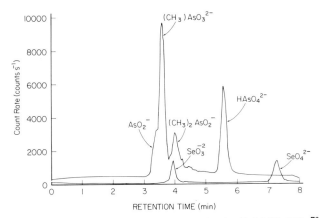

Fig. 5.5. Detection of As and Se species by LC-ICP-MS. ^{78}Se and ^{75}As were monitored. The amount injected of each form was 5 ng (as element, 10 μl of 0.5 mg l^{-1}). Reproduced from *Anal. Chem.* 8 [16] with permission of American Chemical Society.

pairing-reverse phase LC with a mobile phase of 5% methanol in water. It is apparent from the figure that levels of As and Se well below 1 ng can be readily observed. The response of the detector for Se in SeO_3^{2-} is nearly equal to that for Se in SeO_4^{2-}, which is expected because the plasma should atomize these species and ionize the Se from them equally. The response for the various As species is not uniform, which has been attributed to interconversion of As species during sample preparation and separation [16].

Speciation of alkyltin and alkylmercury compounds is of substantial importance for environmental purposes. Separation and selective detection of several alkyl tin species is shown in Fig. 5.6. The separations were performed on a C_{18} column (Vydac G 201 TP, 10 μm) by cation pairing (with $HClO_4$ at 0.023 M) using 5% I methanol at 1.6 ml min^{-1} (for Hg) or 10% methanol at 1.8 ml min^{-1} (for Sn). Monomethyltin and dimethyltin elute together close to the void volume (Fig. 5.6a). Diethyltin is readily separated from the two methyl species. A similar separation for alkylmercury species is shown in Fig. 5.6b. Both these chromatograms show a high background for tin and mercury, which is likely due to bleeding of species containing tin and mercury from the

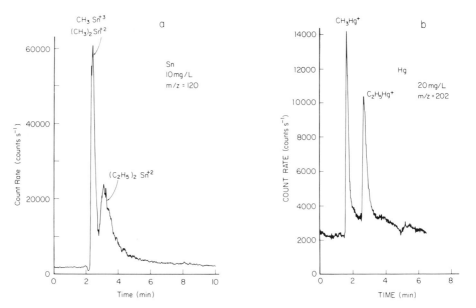

Fig. 5.6. Detection of alkyltin (a) and alkylmercury (b) species. Each tin species was injected at 1 μg (as Sn), each Hg species at 2 μg (as Hg). See text for the separation conditions.

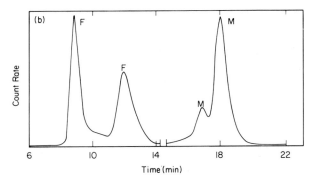

Fig. 5.7. Detection of Cd (m/z = 114) in ferritin (F) and metallothionein (M), separation by size exclusion chromatography. Reproduced from *J. Anal. Atomic Spectrom.* [17] with permission of Royal Society of Chemistry.

column or to memory for these elements in the desolvation system for the nebulizer. If the background in these LC experiments can be reduced to a level closer to that observed usually in ICP-MS ($\sim$10 counts s^{-1}), prospects are good for sensitive detection of these species, for the sensitivity for tin and mercury is 5000–10000 counts s^{-1} per μg of metal injected.

Selective detection of metals in proteins is another potential application of LC-ICP-MS. An example from a recent publication by Ebdon and co-workers is given in Fig. 5.7. Ferritin and metallothionein are separated by size exclusion while the ICP-MS device monitors cadmium. In ferritin, the cadmium is easily observable although present as only a trace impurity [17].

Most LC-ICP-MS studies have focused on analytes containing metals, as these elements are generally ionized quite efficiently by the plasma. Some nonmetals can be monitored with reasonable results by ICP-MS as well. Jiang and Houk used ion pairing reverse phase separations and monitored ^{31}P and ^{34}S in various compounds. Typical chromatograms are shown in Fig. 5.8. In these cases, the baseline is elevated because of background ions at the same nominal m/z as P^+ and S^+. Nevertheless, detection limits for these elements by ICP-MS are reasonable and are significantly better than those obtained by ICP emission spectrometry (Table 5.1) [32].

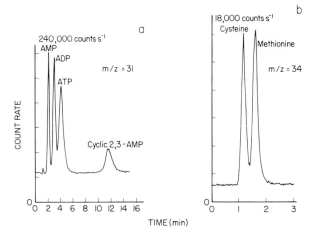

Fig. 5.8 Detection of nonmetals with LC-ICP-MS: (a) P-selective chromatogram for nucleotides, each present at following concentrations: AMP (5 mg P l^{-1}), ADP (8 mg l^{-1}), ATP (11 mg l^{-1}), cyclic-2,3-AMP (5 mg l^{-1}). (b) S-selective chromatogram for sulfur-containing amino acids, each present at 20 mg S l^{-1}. The injection volume was 50 μl. Reproduced from [32] with permission of Pergamon.

TABLE 5.1

DETECTION LIMITS FOR P AND S, 50 μl INJECTIONS [32]

Compound	Detection limit (ng P)	
	LC-ICP-MS	LC-ICP emission
PO_4^{3-}	0.4	78[a], 160[b], 750[c], 500[d]
$P_2O_7^{4-}$	0.6	1000[d]
$P_3O_{10}^{5-}$	1	3000[d]
AMP	0.8	74[a], 750[c]
ADP	1.5	84[a]
ATP	2	92[a]
Cyclic 2,3,-AMP	4	–

Compound	Detection limit (ng S)	
	LC-ICP-MS	LC-ICP emission
SO_4^{2-}	7	16[e]
$S_2O_3^{2-}$	8	–
Cysteine, methionine	6	100–300[f]

References: [a] [33]; [b] [34]; [c] [35]; [d] [36]; [e] [23]; [f] [37].

5.3.2. Applications

Even at this early stage in the development of LC-ICP-MS, there have already been several significant application papers published. Several studies have approached determination of alkylmetal compounds in samples of environmental interest. Beauchemin et al. separated and quantified As species in a fish muscle reference material. Ion pairing with reversed phase chromatography was susceptible to interferences in the separation caused by large excesses of matrix ions (e.g., Fe^{3+}). Ion exchange was found to be superior for separation of As species in that the sample clean-up procedures (e.g., extractions) did not need to be as extensive [38]. Chromatograms from the sample (a) and from a spiked sample (b) are shown in Fig. 5.9. Most of the As is present as arsenobetaine, a non-toxic metabolite of As. The detection limit is 300 pg As [39]. Caruso and co-workers determined various As species in urine by anion exchange. Detection limits were 30–90 pg. Dimethylarsinate, monomethylarsinate, As^{3+} and As^{5+} were quantified at levels of 0.7 to 500 μg l^{-1} in several standard urine samples [40].

Bushee has determined methylmercury at $\sim$870 μg l^{-1} in a different fish reference material and thimerosol in contact lens solutions with good agreement with values reported previously [41]. Detection limits

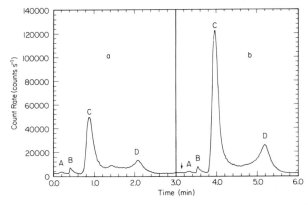

Fig. 5.9. Detection of As (m/z = 75) as several species from fish tissue (reference material DORM-1). Peak A is As(III) + As(V) + monomethylarsonic acid; B is dimethylarsonic acid, C and D are arsenobetaine. (a) is from the sample, (b) is from a sample spiked with 0.14 μg of As as arsenobetaine. The separation is by reversed phase LC with ion pairing. Reproduced from *Anal. Chem.* [39] with permission of American Chemical Society.

References pp. 120–122

112

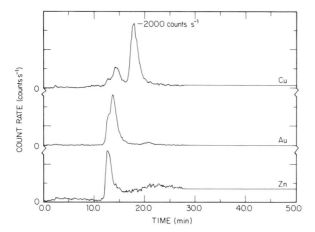

Fig. 5.10. Selected ion chromatograms for Cu, Au and Zn in blood serum from a human patient being treated with auranofin. The separation is by anion exchange. Reproduced from *J. Anal. Atomic Spectrom.* [19] with permission of Royal Society of Chemistry.

for mercury are 60–120 pg using a cold-vapor method for introducing mercury into the plasma.

Elder and co-workers have been investigating LC-ICP-MS for determining metabolites of gold drugs used in arthritis therapy. Selected ion chromatograms for Cu, Au and Zn in blood serum from a patient on auranofin therapy are shown in Fig. 5.10. There are apparently three different Au containing species in this particular sample; the two largest peaks likely represent Au bound to human serum albumin. The total Au level is in the range 300–500 μg l^{-1} in the patient's blood [19]. The distribution of Au, Cu and Zn varies with time on therapy [42]; these three metals could be monitored in the same sample injection because of the multielement capability of ICP-MS.

Mason and co-workers are studying metal binding to proteins and metal detoxification in marine organisms by size exclusion chromatography with ICP-MS. The organisms (polychaetes) were grown in seawater spiked with Cd, and metallothionein was shown to bind the Cd and a variety of other metals as well. Apparently, this was a significant mechanism by which these organisms detoxify absorbed trace metals. Again, multielement chromatograms led to some interesting observations. For example, exposing the polychaetes to Cd also induced a deficiency of Cu. Despite their chemical similarity, Cd and Zn were associated with different suites of metal-binding ligands. Synthesis of metallothionein by the polychaetes was induced by exposure to Cd but not Zn. Thus, the multielement capabilities of ICP-MS facilitated studies of synergistic or antagonistic effects between different metals [18, 43, 44]. This method should prove valuable for study of the molecular and cellular mechanisms of metal homeostasis.

Crews et al. have applied their size-exclusion separations to a study of Cd speciation in pig kidney. Cadmium in different proteins could be distinguished in cooked, uncooked, and digested pig kidney. The bulk of the soluble cadmium in retail pig kidney was associated with a metallothionein-like protein, some of which survived cooking and digestion. The detection limits of ICP-MS were sufficient to permit studies of Cd speciation at "normal" levels of Cd; previous work on this subject by other methods was generally restricted to abnormally high levels of Cd [45].

5.4. CHROMATOGRAPHIC PROCEDURES FOR ALLEVIATING INTERFERENCES

In ICP-MS, the presence of a concentrated sample matrix can cause two general types of interferences. First, polyatomic ions from the matrix element(s) (e.g., MO^+), although generally 1% or less relative to the monatomic matrix ions, can overlap with trace analyte ion(s) of interest. Procedures for correcting for such a spectral overlap are straightforward but are of doubtful accuracy if the polyatomic ions from the matrix are intense compared to the analyte ions. The second type of interference is indicated by a dependence of the analyte ion signal on the identity and total concentration of matrix element(s). Generally, the analyte ion signal is depressed in the presence of a matrix [27–30], although some enhancements have been reported [30]. The effect is most severe when the analyte is light and the matrix is heavy, so that the most difficult analysis is determination of trace Li in uranium [28, 46]. This interference generally becomes evident at matrix concentrations below those that cause significant plugging of the orifice and generally is the limiting factor in the total solute level that can be tolerated.

Chromatographic separations can be used to alleviate many instances of the above interferences. There are two general approaches to such separations. First, a column can be derivatized with a chelating agent (e.g., 8-hydroxyquinoline) that retains the analytes and passes the interferent(s). The eluent is then changed to remove the analytes and allow them to reach the ICP. Thus, the interfering matrix is removed, and some preconcentration can be achieved as well. This approach has been evaluated by McLaren and Beauchemin with excellent results with the column either on-line or off-line [47–49].

The other approach, investigated by the author's group in collaboration with J.S. Fritz, is to add the complexing agent to the sample first and then allow the complexes to be retained on the column. Here, the complexing reagent [bis(carboxymethyl) dithiocarbamate] reacts with a variety of analyte ions but not with either alkali metals or alkaline earth metals. The trace analyte complexes are retained on the column and can be removed later by simply changing to a more basic eluent. In Fig. 5.11, results for chromium, nickel and cobalt in a seawater standard reference material are shown. After the eluent is changed (indicated by the arrow), the trace metals, including cobalt at only ~20 ng l^{-1}, are easily observed and quantified. The seawater was analyzed with very little sample preparation other than simple addition of the reagent [50]. Direct analysis of seawater with present ICP-MS devices would require at least a ten-fold dilution, with a similar degradation of detection limits.

Use of external complexing reagents also permits complexation and retention of the interfering ions. An example is shown in Fig. 5.12. In this case, the objective is to determine trace Cd in a large excess of Mo. The oxide ions MoO$^+$ obscure all the useful isotopes of Cd if the sample is analyzed directly [52]. To circumvent this problem,

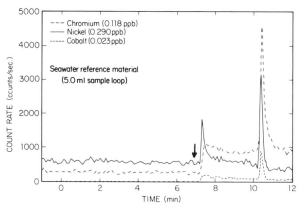

Fig. 5.11. Preconcentration and elution of trace metals in seawater (reference material (CASS-1). Concentrations injected are 0.094, 0.23, and 0.018 μg l^{-1} for Cr, Ni, and Co. Legend lists concentrations in original sample. The sample was injected at time zero, and the eluent was changed to remove the analyte complexes at the time indicated by the arrow. Reproduced from *Anal. Chem.* [50] with permission of American Chemical Society.

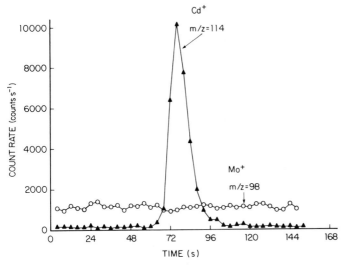

Fig. 5.12. Multiple ion chromatogram for single injection of solution containing Cd (0.1 mg l^{-1}) and Mo (100 mg l^{-1}). The Mo is complexed by N-methylfurohydroxamic acid and is retained by the column. Reproduced from *Anal. Chim. Acta* [51] with permission of Elsevier.

N-methylfurohydroxamic acid is added to complex the Mo. The Mo complex is retained efficiently on XAD resin while the Cd simply elutes through with the void volume in a fashion analogous to flow injection. The molar ratio of Mo/Cd is 1000/1 in this case, yet just a little Mo escapes the column (hence the background at $m/z = 98$). This slight leakage of Mo is not enough for MoO$^+$ to be a problem at the m/z values used for determination of Cd. This same complexing agent is useful for retention of many other ions in which the metals are present in oxidation states of 4+ or higher. Interestingly, many of these metals either cause significant oxide interferences (e.g., Ti and Zr) or induce substantial suppression of analyte signals (e.g., U). The analyte band is, of course, diluted by dispersion in the lines and column, so detection limits are sacrificed somewhat by this approach.

5.5. OTHER SEPARATION METHODS WITH ICP-MS

5.5.1. Solvent extraction

Many of the LC-ICP-MS studies mentioned above employ solvent extraction for fractionating and cleaning up the sample prior to the analytical separation. Solvent extraction has other uses with ICP-MS, three of which will be described below.

ICP-MS is particularly valuable for isotope ratio measurements to support large-scale isotope tracing studies of 7 elemental pathways in biological organisms and in the environment. Most of the ICP-MS methods for isotope ratio measurements in biological samples employ some separation scheme to remove the analyte from the matrix. For example, Serfass and co-workers commonly extract Zn as diethyldithiocarbamate complexes into tetrachloromethane and then back into aqueous acids prior to analysis [53]. Janghorbani's group has described numerous separation methods for various elements [54–56]. These separations are done both to remove interferents and to optimize the solution concentration of the analyte so that the isotope ratio can be measured with the desired precision. Isotope tracer studies involving human or animal subjects are expensive and time-consuming. The resulting samples are literally priceless. Chemical separations take additional time and effort but are desirable if they ensure that the measured isotope ratios will suffice for the purposes of the study.

Beauchemin and co-workers have described an extraction procedure for determining alkyl mercury species in fish and lobster tissue. Organomercury was extracted as the chloride into toluene to isolate it from inorganic mercury. The organomercury was then extracted back into an aqueous medium containing cysteine acetate. Flow injection and isotope dilution were employed to minimize matrix effects and orifice clogging caused by the high salt levels (>4% NaCl) in the final sample. Detection limits for mercury were 2 to 5 μg l^{-1}, and recoveries: were 75 to 100%. Methylmercury was found to be the only significant organomercury, and the methylmercury comprised a substantial fraction (40–90%) of the total mercury in these 6 samples. As shown in Table 5.2, the concentrations of methylmercury determined by ICP-MS agreed well with values obtained by other methods and with the recommended values for these standard reference materials [57]. Thus, the combination of solvent extraction with ICP-MS provides a viable approach for an important environmental problem, namely that of mercury speciation.

Determination of trace impurities in uranium is one of the potentially major applications of ICP-MS because the mass spectrum of uranium from an ICP is very simple. Spectral overlap with analyte signals is not a problem, as is often the case in many optical emission methods. In ICP-MS, however, this analysis is often limited by suppression of analyte signal induced by the large excess of uranium in the solution. Various selective extracting agents (e.g, tri-n-octylphosphine oxide) have been

TABLE 5.2

DETERMINATION OF METHYLMERCURY IN FISH AND LOBSTER TISSUE [57]

Technique	Methylmercury level (μg g^{-1})	
	DORM-1[a]	TORT-1[b]
ICP-MS	0.721 ± 0.033	0.13 ± 0.02[c]
GC with electron capture	0.730 ± 0.035	0.13 ± 0.03
Cold vapor atomic absorption	0.728 ± 0.061	–
Recommended value	0.74 ± 0.06	0.13 ± 0.02[d]

[a] Dogfish muscle tissue. Standard reference material from National Research Council of Canada.
[b] Lobster hepatopancreas. Also from National Research Council of Canada.
[c] Mean (n = 3–10) ± standard deviation.
[d] Uncertainty is 95% confidence level.

employed over the years for removing uranium. Palmieri et al. have described a procedure based on N,N-dihexylacetamide that removes uranium efficiently but leaves virtually all analytes (except thorium) behind. Rare earths and transition metals were quantified by ICP-MS in uranyl nitrate in this fashion [58].

5.5.2. Gas chromatography

Element-selective detection for gas chromatography (GC) based on atomic emission from a plasma has been around on a research basis for some time. The recent introduction of several commercial plasma emission detectors (e.g., by Cetac and Hewlett-Packard) has stimulated renewed interest in this idea.

ICP-MS can be used as such an element-selective detector for GC, as shown in two publications by Van Loon and co-workers [59], and Chong and Houk [60]. To illustrate and evaluate this concept, both groups used rather simple separations on a packed column with a heated metal transfer line to transfer the eluted compounds to the plasma. A selective chromatogram for P is depicted in Fig. 5.13. Detection limits vary depending on the degree of ionization of the element monitored and the background for that element. Elements such as Sn, P and I are ionized with reasonable efficiency and have detection limits of 0.1 to 10 ng. Carbon and oxygen ions can be observed, but the detection limits are compromised by low ionization efficiencies and by the substantial levels of these elements as impurities in the argon used for the plasma. Isotope ratios can be determined using elemental ions in individual

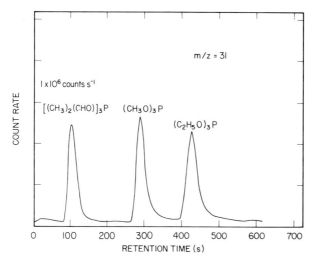

Fig. 5.13. GC-ICP-MS: P$^+$-selective chromatogram of phosphorous esters, equal volumes of each compound injected. Note similar response for P in the three compounds. Reproduced from *Appl. Spectrosc.* [60] with permission of Society for Applied Spectroscopy.

compounds, as shown by the spectra for sulfur and bromine in Fig. 5.14.

These early studies illustrated the essential feasibility of combining GC separations with ICP-MS. Further improvements in this technique are likely for several reasons. First, the early work was performed using prototype ICP-MS devices. The newer instruments perform substantially better than these old stalwarts, particularly with respect to the

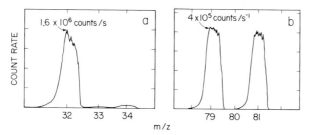

Fig. 5.14. GC-ICP-MS: Mass spectra showing isotope peaks of atomic ions: (a) S$^+$ from 1.1 mg of thiophenol (0.31 mg S); (b) Br$^+$ from 1.5 mg of bromoethane (1.1 mg Br). Reproduced from *Appl. Spectrosc.* [60] with permission of Society for Applied Spectroscopy.

background and efficiency of ion transmission. Second, the use of capillary columns, as reported by Caruso and co-workers [61], will certainly improve the quality of the separations. The development of an interface with very low dead volume [62] should facilitate use of capillary columns with GC-ICP-MS. Finally, there is substantial interest in MS with either ICPs or microwave plasmas sustained in helium [61, 63–67]. The ionization energy of helium (24.6 eV) is substantially greater than that of argon (15.76 eV) so larger populations of more energetic electrons are likely in a helium plasma. Thus, helium plasmas are generally superior to argon ICPs at exciting emission from nonmetals, particularly in GC applications where there is no solvent load and the plasma need not be as robust as for introduction of solution aerosols. Such helium plasmas may prove superior at generating ions from nonmetals for MS as well. Recent work with helium plasmas is described by J.A. Caruso in this volume.

5.6. CONCLUSIONS

The studies cited have shown that LC-ICP-MS has the powers of detection and selectivity necessary for elemental speciation in several real analytical situations. Each of these studies also points out that further improvements in the separations, background levels, and memory are also desirable. Most of the applications studied so far have involved rather high levels of the analytes, so further improvements in powers of detection are desirable. Sample preparation and chromatographic procedures need to be carefully developed so that the elemental species present are not perturbed chemically. Some potentially useful separation procedures have been devised for alleviating interferences, but these have seen little routine use as yet in the analytical community. Such improvements will doubtless extend the scope of LC-ICP-MS to a wider range of real applications. Growth is expected for GC-ICP-MS as well, although there is presently less interest and research in this area than in LC.

ACKNOWLEDGEMENT

Ames Laboratory is operated for the U. S. Department of Energy by Iowa State University under Contract No. W-7405-ENG-82. This research was supported by the Office of Basic Energy Sciences.

REFERENCES

1 R.S. Houk, V.A. Fassel, G.D. Flesch, H.J. Svec, A.L. Gray and C.E. Taylor, Anal. Chem., 52 (1980) 2283–2289.
2 D.W. Kopenaal, Anal. Chem., 60 (1988) 113R–131R.
3 V.A. Fassel, Science, 202 (1978) 183–191.
4 R.M. Barnes, CRC Crit. Reviews Anal. Chem., 7 (1978) 203–296.
5 R.S. Houk, Anal. Chem., 58 (1986) 97A–105A.
6 R.S. Houk and J.J. Thompson, Mass Spectrom. Reviews, 7 (1988) 425–462.
7 A.L. Gray, Spectrochim. Acta, Part B, 40B (1985) 1525–1537.
8 A.R. Date and A.L. Gray, (Eds.), Applications of Inductively Coupled Plasma Mass Spectrometry, Blackie, London, 1988.
9 J.A. Olivares and R.S. Houk, Anal. Chem., 57 (1985) 2674–2679.
10 D.J. Douglas and J.B. French, J. Anal. Atomic Spectrom. 3 (1988) 743–747.
11 D.J. Douglas and L.A. Kerr, J. Anal. Atomic Spectrom. 3 (1988) 749–752.
12 N. Bradshaw, P.D. Blair, D. Kirby, Pittsburgh Conf. on Anal. Chem. Appl. Spectrosc., Atlanta, GA, March 6–10, 1989, Paper 1274.
13 R.S. Houk, in K.A. Gschneidner and L. Eyring (Eds.), D Handbook of Phys and Chem. of Rare Earths, Vol. 13, North-Holland Physics Publishing, Amsterdam, (1989) in press.
14 G. Zhu and R.F. Browner, J. Anal. Atomic Spectrom., 3 (1988) 781–789.
15 R.C. Hutton and A.N. Eaton, J. Anal. Atomic Spectrom. 2 (1987) 595–598.
16 J.J. Thompson and R.S. Houk, Anal. Chem., 58 (1986) 2541–2548.
17 J.R. Dean, S. Munro, L. Ebdon, H.M. Crews and R.C. Massey, J. Anal. Atomic Spectrom., 2 (1987) 607–610.
18 A.Z. Mason, S.D. Storms and K. Jenkins, Anal. Biochem. (1989) submitted.
19 S.G. Matz, R.C. Elder and K. Tepperman, J. Anal. Atomic Spectrom., (1989) accepted.
20 K.W. Olson, W.J. Haas, Jr. and V.A. Fassel, Anal. Chem. 49 (1977) 632–637.
21 B.R. Bear and V.A. Fassel, Spectrochim. Acta, Part B, 41B (1986) 1089–1113.
22 K.E. Lawrence, G.W. Rice and V.A. Fassel, Anal. Chem., 56 (1984) 289–292.
23 K.E. LaFreniere, V.A. Fassel and D.E. Eckels, Anal. and Chem., 59 (1987) 879–887.
24 A.L. Gray, R.S. Houk and J.G. Williams, J. Anal. Atomic Spectrom., 2 (1987) 13–20.
25 G. Zhu and R.F. Browner, Appl. Spectrosc. 41 (1987) 349– 359.
26 M.-A. Vaughn, G. Horlick and S.H. Tan, J. Anal. Atomic Spectrom., 2 (1987) 765–772.
27 J.S. Crain, R.S. Houk and F.G. Smith, Spectrochim. Acta, Part B, 43B (1988) 1355–1364.
28 S.H. Tan and G. Horlick, J. Anal. Atomic Spectrom., 2 (1987) 745–763.
29 C. Vandecasteele, M. Nagels, H. Vanhoe and R. Dams, Anal. Chim. Acta, 211 (1988) 91–98.
30 D. Beauchemin, J.W. McLaren and S.S. Berman, Spectrochim. Acta, Part B, 42B (1987) 467–490.
31 W.B. Jones, R.C. Elder and K. Tepperman, Pittsburgh Conf. on Anal. Chem. Appl. Spectrosc., Atlanta, GA, March 6–10, 1989, Paper 1377.
32 S.-J. Jiang and R.S. Houk, Spectrochim. Acta, Part B, 43B (1988) 405–411.
33 K. Yoshida, H. Haraguchi and K. Fuwa, Anal. Chem., 55 (1983) 1009–1012.
34 W.R. Biggs, J.T. Gano and R.T. Brown, Anal. Chem., 56 (1984) 2653–2657.
35 D.R. Heine, M.B. Denton and T.D. Schlaback, Anal. Chem., 54 (1982) 81–84.

36   M. Morita and T. Uehiro, Anal. Chem., 53 (1981) 1997–2000.
37   K. Yoshida, T. Hasegawa and H. Haraguchi, Anal. Chem., 55 (1983) 2106–2108.
38   D. Beauchemin, K.W.M. Siu, J.W. McLaren and S.S. Berman, J. Anal. Atomic Spectrom., 4 (1989) 285–289.
39   D. Beauchemin, M.E. Bednas, S.S. Berman, J.W. McLaren, K.W.M. Siu and R. E. Sturgeon, Anal. Chem., 60 (1988) 2209–2212.
40   D. Heitkemper, J. Creed, J. Caruso and F.L. Fricke, J. 6 Anal. Atomic Spectrom., 4 (1989) 279–284.
41   D.S. Bushee, Analyst, 113 (1988) 1167–1170.
42   R.C. Elder, K. Tepperman, M.L. Tarver, S. Matz, W.B. Jones and E.V. Hess, Arthritis and Rheumatism, (1989) submitted.
43   A.Z. Mason, S.D. Storms and K. Jenkins, Mar. Environ. # Res. (1989) submitted.
44   A.Z. Mason, S.R. Howe and K. Jenkins, Winter Conf. on E Plasma Spectrochemistry, San Diego, CA, January 3–9, 1988, Paper S18.
45   H.M. Crews, J.R. Dean, L. Ebdon and R.C. Massey, Analyst, 114 (1989) 895–899.
46   G.R. Gillson, D.J. Douglas, J.E. Fulford, K.W. Halligan and S.D. Tanner, Anal. Chem., 60 (1988) 1472–1474.
47   J.W. McLaren, A.P. Mykytiuk, S.N. Willie and S.S. Berman, Anal. Chem., 57 (1985) 2907–2911.
48   D. Beauchemin, J.W. McLaren, A.P. Mykytiuk, and S.S. Berman, J. Anal. Atomic Spectrom., 3 (1988) 305–308.
49   D. Beauchemin and S.S. Berman, Anal. Chem., 61 (1989) 1857–1862.
50   M.R. Plantz, J.S. Fritz, F.G. Smith and R.S. Houk, Anal. Chem., 61 (1989) 149–153.
51   S.-J. Jiang, M.D. Palmieri, J.S. Fritz and R.S. Houk, Anal. Chim. Acta, 200 (1987) 559–571.
52   C.W. McLeod, A.R. Date and Y.Y. Cheung, Spectrochim. Acta, Part B, 41B (1986) 169–174.
53   R.E. Serfass, J.J. Thompson and R.S. Houk, Anal. Chim. Acta, 188 (1986) 73–84.
54   M. Janghorbani, in A.R. Date and A.L. Gray (Eds.), Applications of Inductively Coupled Plasma Mass: Spectrometry, Blackie, London, 1988, Ch. 8.
55   X.F. Sun, B.T.G. Ting, S.H. Zeisel and M. Janghorbani, Analyst, 112 (1987) 1223–1228.
56   S. Schuette, D. Vereault, B.T.G. Ting and M. Janghorbani, Analyst, 113 (1988) 1837–1842.
57   D. Beauchemin, K.W.M. Siu and S.S. Berman, Anal. Chem., 60 (1988) 2587–2590.
58   M.D. Palmieri, J.S. Fritz, J.J. Thompson and R.S. Houk, Anal. Chim. Acta, 184 (1986) 187–196.
59   J.C. Van Loon, L.R. Alcock, W.H. Pinchin and J.B. French, Spectrosc. Letters, 19 (1986) 1125–1135.
60   N.S. Chong and R.S. Houk, Appl. Spectrosc., 41 (1987) 66–74.
61   A.H. Mohamad, J.T. Creed, T.M. Davidson and J.A. Caruso, Appl. Spectrosc. 43 (1989) 1127–1131.
62   B.R. LaFreniere, R.S. Houk, D.R. Wiederin and V.A. Fassel, Anal. Chem., 60 (1988) 23–26.
63   A. Montaser, S.K. Chan and D.W. Koppenaal, Anal. Chem., 59 (1987) 1240–1242.
64   D.W. Koppenaal and L.F. Quinton, J. Anal. Atomic Spectrom., 3 (1988) 667–672.
65   R.D. Satzger, F.L. Fricke, P.G. Brown and J.A. Caruso, Spectrochim. Acta, Part B, 42B (1987) 705–712.

66 P.G. Brown, T.M. Davidson and J.A. Caruso, J. Anal. Atomic Spectrom., 3 (1988) 763–769.

67 J.T. Creed, A.H. Mohamad, T.M. Davidson, G. Ataman and J.A. Caruso, J. Anal. Atomic Spectrom. 3 (1988) 923–926.

I.S. Krull (Ed.), *Trace Metal Analysis and Speciation*
Journal of Chromatography Library Series, Vol. 47
© 1991 Elsevier Science Publishers B.V., Amsterdam

Chapter 6

The Use of Complexing Eluents for the High Performance Liquid Chromatographic Determination of Metal Species

B.D. KARCHER * and I.S. KRULL
Barnett Institute of Chemical Analysis and Materials Science, 341 Mugar Building, Northeastern University, Boston, MA, 02115, U.S.A.

6.1. INTRODUCTION

Within the past ten years numerous advances have been realized for the determination of metal species using high performance liquid chromatographic (HPLC) methods of analysis. Separation efficiencies have increased, detection sensitivities improved, and separations for samples of greater diversity and complexity have been accomplished. These achievements have resulted from new and/or improved column packings, innovative detectors, more efficient and chemically inert hardware, and the willingness on the part of the analyst to investigate and apply different chemistries in developing new methods of analysis. Some of the more successful of these methods are those which are based upon the use of multidentate complexing agents, i.e., chelating agents. The numerous and varied chromatographic applications of these reagents is evidence that *chelation reactions* have played a paramount role in the evolution of HPLC analyses for metal species. In a more general sense, however, it should be realized that nearly all of the liquid chromatographic methods which have been used for separating metal species are, by definition, *complexation reactions*.

The term *complexation* refers to the reaction of a metal ion with an electron donor group (ligand), and entails the transformation of a solvated metal ion into a complex ion. In aqueous solutions of metal salts, the metal cations are surrounded by a sheath of aquo groups which are collectively called the primary hydration sphere. These solvated metal ions are not referred to as complex ions. Instead, a complex ion is formed if one or more of the solvent molecules within the primary hydration sphere of the metal ion is replaced either by ions, or by

---

*Present affiliation: Bristol-Myers Squibb, One Squibb Drive, P.O. Box 191, New Brunswick, NJ 08903-0191, U.S.A.

References pp. 165–166

molecules other than solvent molecules. The complexation of a metal ion in aqueous solution can be expressed by the following equation:

$$M(H_2O)_n^m + L^l \rightleftharpoons ML(H_2O)_{n-1}^{m-l} + H_2O \tag{1}$$

where M denotes the metal ion and m is its charge, L represents any ligand (charged group or molecule) other than water, and l is the charge of the ligand. The remaining aquo groups in the complex may be successively replaced by other L groups to form the complex ML_n, where n indicates the maximum coordination number of the metal ion.

The fact that several metal ions have hydration energies of several thousand kilojoules indicates that the solvent sheath is an essential facet of a metal ion in solution [1]. The composition of the solvent sheath—whether other ligands have replaced solvent molecules within the solvent sheath—determines whether a complex ion has been formed. Aside from those reactions which can be described as purely electro-chemical in nature, the remaining aqueous chemistry of metal ions is based largely on complexation chemistry. Even the classic ion-exchange separations on sulfonated exchange resins with strong acid eluents are considered complexation reactions. This is because during the elution process, the aquated metal analyte ions undergo continuous exchange reactions with the following species as the band of metal ions traverse the column: water, sulphonate groups of the stationary phase, and the negative charged conjugate base of the acid eluent. Eventually, the metal elutes from the column in a more or less *complexed* state.

Because the term *complexation* is rather broad in scope, we indicate without reservation that our intention is not to exhaustively review each *complexation*-based liquid chromatographic system that has been used for the separation of metal-species: we won't even propose an attempt at all of the ion-exchange methods. Rather, we intend for this treatise to serve as a review and discussion of recent publications (1979–1989) pertaining to the HPLC separation of metal species using eluents containing multidentate complexing agents. Throughout this chapter we use the term *complexation* to refer to the reaction of a metal ion with a multidentate complexing agent.

The use of complexing agents in HPLC analyses are numerous and varied. One method which has been used to classify the different techniques is based on *when* the complexing agent is reacted with the metal ion. There are pre-column and post-column techniques, along with *in situ*, which refers to derivatization within the column when the complexing agent is a component of the mobile phase. These three techniques have offered the analyst profoundly different analytical capabilities. For instance, the primary utilization of pre-column methods has been to derivatize metal ions to form hydrophobic metal chelates which can be separated using reverse-phase columns and aqueous-organic eluents. This approach became popular rather quickly within the chromatography community because it represented a major inroad to the use of reverse-phase technology to separate polar, inorganic analytes.

In situ derivatization, on the other hand, has been widely utilized in order to elute metal ions from either cation-exchange or dynamically coated reverse-phase columns. In addition to an increase in column efficiency over traditional single- or dual-column cation-exchange separations, the *in situ* use of complexing agents has facilitated polyvalent cation separations without the need of concentrated eluent solutions. In fact, a direct outcome of using complexing agents in the mobile phase has been the development of a method for the analysis of polyvalent metal ions which entails single-column cation-exchange separation with conductivity detection [2].

As for post-column reactions, it is widely recognized that post-column techniques are useful for improving the detection sensitivity for both organic and inorganic analytes. Improved detection of inorganic species has been realized because the post-column chemistries are designed to produce reaction products which can be quantified using either absorbance, fluorescence, or electrochemical detection. In addition to the advantages in sensitivity and selectivity apparent with the solute property detectors, there is also an added bonus because these detectors have fewer and less critical constraints with respect to the composition and concentration of the mobile phase. As a result, numerous innovative and efficient separations have been developed for inorganic samples. One such example is the analysis of transition metal ions. Until just a few years ago, transition metal ions were separated using dual-column cation-exchange chromatography, and quantified using conductivity detection. Unfortunately, the concentrated eluents that were necessary to elute these ions often resulted in low chromatographic efficiency and/or selectivity. In addition, this method required frequent suppressor column regeneration, and metal-hydroxide precipitates were often formed when hydroxide-containing regenerant solutions were used. A more satisfactory method for the analysis of transition metals involves a cation-exchange separation, followed by post-column complexation with 4-(2-pyridylazo)resorcinol (PAR) to form chelates which can be sensitively monitored using absorption detection [3].

In spite of the obvious impact that pre-column and/or post-column use of complexing agents has had on advancing the field of HPLC separations for metal species, neither technique has been reviewed in this chapter. The reason for this is that the scope of pre-column complexation reactions is generally very narrow; powerful complexing agents are used to form stable metal complexes which are then separated using either reverse-phase or normal-phase chromatography. Post-column complexation reactions, on the other hand, often emphasize the detection aspect of metal ion analyses, whereas the emphasis of this chapter is on separations. The many innovative and efficient separations which have been achieved for inorganic species using complexing agents *in situ* has drawn our attention to this important class of derivatization reaction. Readers interested in either pre- or post-column derivatization techniques are directed to a recent review of these subjects [4].

The first part of this chapter is a review of the literature. It was written with the intent to summarize the important contributions that *in situ* derivatizations using multidentate complexing agents have made to the development of HPLC methods of analysis for metal species. The authors have not presented an exhaustive review of the literature. Rather, special consideration has been extended to those publications which describe: innovative methods for separating metal cations, important applications, or studies concerned with elucidating retention mechanisms. The second and final part of this chapter deals with the mathematical language of complex equilibria and its implementation in designing separations for difficult samples of metal species.

6.2. COMPLEXING ELUENTS

Those citations reviewed in this treatise have been categorized into subsections according to the particular complexing agent used in the study. The subsections are arranged in alphabetical order using the name (common name when possible) of the complexing agent, and within each subsection the papers are discussed in chronological order, from past to present. Studies which have utilized eluents containing more than one complexing agent have been categorized within subsection 6.2.11, entitled "Mixed Complexing Eluents". The structures for some of the more commonly used complexing agents are presented in Fig. 6.1.

6.2.1. Acetic acid

Blaszkewicz and Neidhart [5] investigated the HPLC determination of inorganic lead and some organolead species of concern to occupational

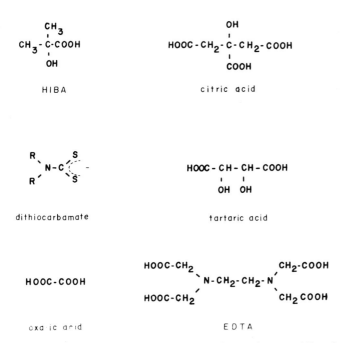

Fig. 6.1. Complexing agents commonly used as mobile phase eluents. HIBA = α-hydroxyisobutyric acid, EDTA = ethylenediaminetetra-acetic acid, R=alkyl or aromatic group.

health, the environment, and industrial toxicology. The lead species analyzed included Pb(II) nitrate, trimethyllead acetate, triethyllead chloride, tetramethyllead, and tetraethyllead. The authors implicate HPLC as the method of choice for the analysis of these lead species due to the rather mild eluting conditions, and the availability of high efficiency column packings.

The authors initially considered two different modes of separation in order to resolve the lead species—ion-exchange on strong-acid cation-exchangers, and reverse-phase—but high concentrations of a strong-acid were needed in order to elute Pb(II), and this caused premature corrosion of the metal components of the HPLC apparatus. As a result, the authors directed their concerted efforts to the development of a reverse-phase method. A 10-μm Li-Chrosorb C_{18} column, 300 mm × 2-mm i.d., was initially tested using mobile phases comprised of varying mix ratios of 0.1 M acetate buffer and methanol, but the five lead species could not be resolved using this column and conditions.

The authors next tested a 5-μm Nucleosil C_{18} column, 200 mm × 4-mm i.d., using similar mobile phases, and all five lead species could

128

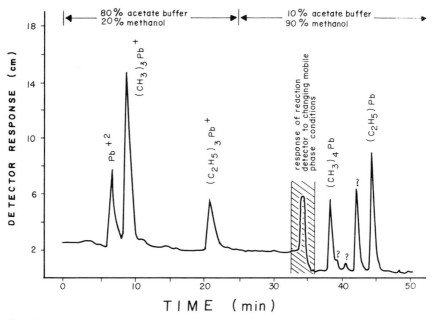

Fig. 6.2. Separation of organolead compounds. Column: 5-μm Nucleosil C_{18}, 200 mm × 4-mm i.d. Mobile phases: #1 = 0.1 M/0.1 M acetate buffer : methanol (80 : 20), pH 4.6; #2 = 0.1 M/0.1 M acetate buffer : methanol (10 : 90), pH 4.6. Column temperature: 21°C. Peaks: Pb(II), (1 μg); Me_3Pb^+, (1 μg); Et_3Pb^+, (3 μg); Me_4Pb, (5 μg); Et_4Pb, (4 μg). Injection volume: 20 μl. Detection: post-column degradation using I_2, followed by reaction with 10^{-4} M PAR for spectrophotometric detection at 546 nm. (Reproduced with permission from ref. [5], copyright 1983.)

be separated within 50 minutes when the mobile phase was changed from 80 : 20 acetate buffer/methanol to 10 : 90 acetate buffer/methanol at 25 minutes into the separation. A chromatogram illustrating the separation of the lead species is reproduced in Fig. 6.2. The 80 : 20 acetate buffer/methanol eluent was used to separate the ionic lead species Pb(II), Me_3Pb^+, and Et_3Pb^+; and the 10 : 90 acetate buffer/methanol eluent was used to elute the tetraalkyllead compounds.

In order to quantify the metal species, the authors sought to post-column react the lead species with 4-(2-pyridylazo)-resorcinol (PAR) in order to form intense red complexes, which could be detected using absorption monitoring. However, because only the dialkyllead compounds and Pb(II) were complexed by PAR, it was necessary to mix the effluent with a dilute solution of iodine prior to the addition of the PAR reagent. This was done to cleave the lead–carbon bonds of the tri- and tetraalkyllead compounds, thereby making these species amenable for complexation with PAR. Each peak was then quantified by measuring the absorbance of the PAR complexes, and the detection limits were in the range of 0.3 to 1.5 nmole for the five lead species.

Langseth [6] also used mobile phases which contained acetate to resolve organometal species, but his studies were directed towards developing a simple HPLC method which could be used for the separation and quantification of trace amounts of diphenyltin and dialkyltin compounds. Langseth self-imposed the limitation to use either pre-column or *in situ* chelation of the organotin species, in order to form complexes that could be quantified using either absorbance or fluorescence detectors. The study involved the chelating agents 2',3,4',5,7-pentahydroxyflavone (morin) and 3-hydroxyflavone, and normal-phase chromatography was selected because of solubility problems with the organotin complexes in aqueous eluents. The organotin-morin complexes proved to be intensely fluorescent, and optimization of the entire system was based on detection sensitivity while vying for the greatest

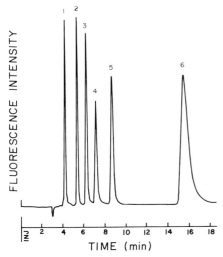

Fig. 6.3. Chromatogram of diphenyl- and dialkyltin compounds. Columns: 30 mm × 4.6-mm i.d. cyanopropyl guard, and 250 mm × 4.6-mm i.d. cyanopropyl analytical (Brownlee Labs). Mobile phase: toluene with 3% acetic acid, 2% methanol, and 0.0015% morin; flow rate 1.0 ml min^{-1}. Peaks: 1 = Oc$_2$SnCl$_2$, (60 ng); 2 = Bu$_2$SnCl$_2$, (60 ng); 3 = Pr$_2$SnCl$_2$, (60 ng); 4 = Ph$_2$SnCl$_2$, (20 ng); 5 = Et$_2$SnCl$_2$, (60 ng); 6 = Me$_2$SnCl$_2$, (120 ng). Detection: fluorescence, 420 nm excitation and 500 nm emission. (Reproduced with permission from ref. [6], copyright 1984, Pergamon Press.)

chromatographic selectivity and resolution. Toluene was found to be the best major component of the mobile phase, and reproducible peak heights were obtained with 0.0015% morin in the eluent. Acetic acid was added to the mobile phase and this proved to be essential in order to optimize chromatographic resolution and selectivity. A chromatogram illustrating the separation of dioctyl-, dibutyl-, dipropyl-, diphenyl-, diethyl-, and dimethyl- tin on a 5-μm cyanopropyl-bonded silica column is reproduced in Fig. 6.3.

Caruso and co-workers [7] separated alkyltin compounds using micellar liquid chromatography and tested the compatibility of this separation technique with inductively coupled plasma mass spectrometric detection (ICP-MS). This study was undertaken because many of the HPLC separations developed for these compounds involve the use of hydro-organic mobile phases, and use of organic solvents with ICP-MS results in a decrease in sensitivity due to excessive solvent loading of the plasma. Separations for many of these species were shown using 5-μm C_{18} Spherisorb silica-bonded columns, 50 mm × 4.6 mm i.d., and mobile phases containing sodium dodecyl sulfate, acetic acid, propanol, and potassium fluoride.

6.2.2. Citric acid

Cassidy and Elchuk [8] investigated HPLC enrichment, separation, and detection of trace concentrations of the divalent metals Ni, Co, Zn, Pb, and Mn. This work is significant in two respects: strong-acid ion-exchange enrichment cartridges were shown to give quantitative recoveries of metal ions from aqueous solutions at the ng ml^{-1} and pg ml^{-1} levels; and simultaneous enrichment and analysis of samples were demonstrated for samples which were backflushed onto an analytical column using complexing agents.

In studies of column efficiency, the chromatographic behavior of the test ions Co(II) and Ni(II) was recorded on strong-acid cation-exchange columns from three different manufacturers (Aminex, Brownlee, and Nucleosil) as a function of the mobile phase concentration of citrate or oxalate. Fig. 6.4 illustrates the effect of complexing agent concentration on column efficiency. The authors indicate that changes in the value of the capacity factors (k')—resulting from changes in the concentration of the eluent—did not contribute significantly to changes in the value of HETP. As an example, k' for Co(II) changed from 7.6 to 0.8 for the Aminex resin with an oxalate eluent (curve 2, Fig. 6.4), but little change was observed in HETP.

ELUENT CONCENTRATION (mol L^{-1})

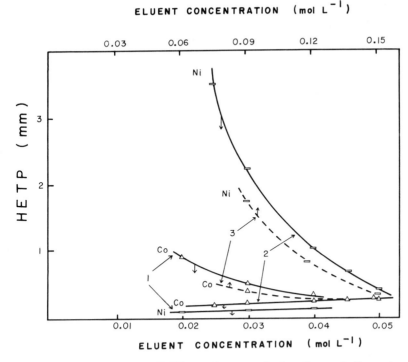

Fig. 6.4. Dependence of HETP on eluent molarity. Curve *1*: Brownlee column with citrate eluent at pH 4.6. Curve *2*: Aminex A-5 column with oxalate eluent at pH 4.6. Curve *3*: Aminex A-5 column with citrate eluent at pH 4.6. Flow rate: 1 ml min^{-1}. (Reproduced from the *Journal of Chromatographic Science* by permission of Preston Publications, a Division of Preston Industries, Inc.)

The relative retention of Co(II) and Ni(II) was the same for each of the three columns. There were, however, indications that the mechanism of ion-exchange was different for each column. For instance, the relative positions of the Co(II) and Ni(II) curves for the Brownlee and Aminex columns using a citrate eluent (curves 1 and 3, Fig. 6.4) are reversed. Although the data for the Nucleosil column were not shown, the authors indicate that the HETP values were constant over the range of citrate concentrations studied.

The primary focus of Kirk and Hewavitharana [9] was to improve the ion-interaction reverse-phase separation of anionic metal species, particularly triply charged species, and to shorten the analysis time. In addition, the authors hoped to broaden the scope of the ion-interaction method for separating cation species to include geometric isomers and

132

some quadruply charged cations of photochemical interest. The authors found that separations could be improved if ions of the same charge as the analytes were added to the mobile phase. For example, separations for anionic analytes were improved if complexing eluents were used.

Using a reverse-phase column and an eluent containing hexanesulfonate to establish the ion-exchange surface, the authors were able to separate the chromium species t-[Cr(tn)$_2$F(NCS)]$^+$, t-[Cr(en)$_2$F(NCS)]$^+$, and t-[Cr(NH$_3$)$_4$F(NCS)]$^+$ by using a methanol gradient. In these formulas (tn) is 1,3-diaminopropane and (en) is 1,2-diaminoethane. Although complexing agents were not used to separate these metal species, by switching to a cationic ion interaction reagent and using otherwise similar chromatographic conditions, the authors attempted to separate anionic complexes of formula Co(CN)$_5$X^{3-}, where X denotes CN$^-$, N$_3^-$, NCS$^-$, Cl$^-$, Br$^-$, and I$^-$. In initial runs using isocratic elution with a mobile phase containing 50% methanol and octylamine as the ion-interaction reagent, the NCS$^-$, Cl$^-$, and Br$^-$ compounds coeluted. The other three species were well resolved, but showed broad and asymmetric peaks. The chain length and nature of the ion-interaction reagent, its concentration, and the methanol/water ratio in the mobile phase were systematically varied in an attempt to improve the separation efficiency of these species. However, even after these parameters had been optimized, the separation efficiency of the anionic complexes could only be marginally improved.

The authors eventually added other anionic compounds to the mobile phase in the hope that peak shapes would improve as a result of competitive interactions. Addition of either fluoride, oxalate, tartrate, citrate, or hexafluorophosphate resulted in improved chromatographic resolution, greater separation factors, improved peak symmetry, and larger theoretical plate counts. The authors note that the most dramatic improvements were recorded for those eluents containing citrate.

6.2.3. Dithiocarbamates

Bond and Wallace [10] described a microprocessor-based chromatographic system which they used for the simultaneous and automated determination of Pb(II), Cd(II), Hg(II), Co(II), Ni(II), and Cu(II). Reverse-phase was used to separate *in situ* formed di-thiocarbamate complexes, and the system could operate continuously and unattended for periods of several days using spectrophotometric detection, and slightly less time using electrochemical detection with background suppression because this mode required frequent suppressor regeneration.

Although the emphasis of this paper was on detection and system automation, the authors utilized mobile phases containing either diethyldithiocarbamate (DEDTC) or pyrollidinecarbodithioate (PYDTC) to separate the six metal ions. Chromatograms were shown for the metals separated on a silica-based C_{18} Spherisorb column using an eluent containing PYDTC, and the only metals that were not baseline resolved were Cd(II) and Hg(II). For each metal studied, the DEDTC complex was retained more strongly than the less hydrophobic PYDTC complex, making a strong case for retention based on partitioning of metal-ligand complexes into the hydro-carbonaceous stationary phase.

Smith et al. [11] focused their attention to resolving some of the chromatographic anomalies documented in one of their earlier publications concerned with the HPLC separation of metal cations using *in situ* derivatization [12]. In that report, this group was unable to achieve good peak shape and reproducible peak areas for Pb(II) and Hg(II) using methanol–water eluents that contained diethyldithiocarbamate (DEDTC). A follow-up study indicated that the poor chromatographic behavior of Pb(II) and Hg(II) was caused from the insolubility of the DEDTC chelates of these metals in the mobile phase. Separations were dramatically improved by the addition of $CHCl_3$ to the mobile phase. Sharp peaks and reproducible peak areas were realized for lead and mercury using a 70:20:10 methanol–water–chloroform mobile phase containing 0.05% sodium DEDTC. However, not all the changes observed were beneficial: the addition of $CHCl_3$ to the mobile phase created non-linearity in the calibration graph for Cd(II). Also, the sensitivity of the method for Cd(II) determination decreased dramatically due to an overlap of the absorbance spectra of the cadmium chelate with the dithiocarbamate reagent.

6.2.4. Ethylenediaminetetraacetic acid (EDTA)

Marina et al. [13] utilized a C_{18} column and an eluent containing EDTA and tetrabutylammonium bromide to separate Pb(II), Mn(II), Cu(II), Ni(II), Cr(III), and Fe(III). The novelty of this work was in the combined use of a chelating agent to form negatively charged metal complexes, and a surface active cationic modifier to retain the metal–EDTA complexes. Absorption of the complexes at 254 nm was used for detection, and the authors indicate detection limits of 100 pg Fe(III), 32 ng Mn(II), and 2 ng Pb(II).

Hagan et al. [14] utilized the powerful complexing ability of EDTA in order to develop an HPLC method to quantify free and chelated gadolinium in drug preparations. Those drugs examined are used as contrast agents for diagnostic applications of nuclear magnetic resonance imaging (MRI). Contrast agents shorten the longitudinal and transverse relaxation time of the water protons in tissues, resulting in enhanced tissue contrast that is dependent on the in vivo distribution of the contrast agent. The most effective paramagnetic ion to be used as a contrast agent has been Gd(III). However, because the Gd(III) aqua ion is relatively toxic, drug formulas usually contain gadolinium complexed by such strong ligands as 1,4,7,10-tetracarboxymethyl-1,4,7,10-tetraazacyclododecane (DOTA) or diethylenetriaminepentaacetic acid (DTPA). The method developed by Hagan et al. consists of separating Gd(III) from the DOTA or DTPA chelated forms of gadolinium on a C_{18} column using an eluent containing 50 mM Tris base and 2 mM EDTA, pH 7.3. Under these conditions, Gd(III) actually eluted as the Gd(EDTA)$^-$ chelate. The fluorescence of gadolinium was used for quantification.

6.2.5. *p*-Hydroxybenzoic acid

Mehra and Frankenberger [15] investigated the potential of single-column ion chromatography with conductivity detection for the determination of molybdate (MoO_4^{2-}) and several non-metal anions common to soils. The authors evaluated the separation of Cl^-, NO_3^-, PO_4^{3-}, SO_4^{2-}, and MoO_4^{2-} using three different anion-exchange columns (Waters polymethacrylate gel IC-Pak, a silica-based 300 IC from Vydac, and a resin-based Wescan anion/R) with several different eluents (sodium hydroxide, borate-gluconate, sodium benzoate, *p*-hydroxybenzoic acid, and potassium hydrogen phthalate). The Vydac silica-based column was found to be unstable within the pH range suited for anionic analyses using single column methodology (pH 7.5 to 9.0). The best resolution, column efficiency, and detection limits were obtained using the Waters IC-Pak column and a *p*-hydroxybenzoic acid eluent. This system is indicative of an anion-exchange mechanism wherein the complexing agent functions as a competing ion.

6.2.6. α-Hydroxyisobutyric acid (HIBA)

Elchuk and Cassidy's [3] investigation of HPLC for the separation and determination of metal ions included a comparison of the performance of 5- and 10-μm bonded phase strong-acid ion-exchangers with a 13-μm styrene-divinylbenzene resin, optimization of post-column reactions using several different strong chelating agents, and on-column preconcentration of 100 ml samples containing 1 ng ml^{-1} of metal cations. The lanthanide metal ions were chosen for the study because these cations have been difficult to separate by liquid chromatography. The eluent in all of the studies contained HIBA, one of the best complexing agents available for the ion-exchange separation of the lanthanides [16].

The authors studied the influence of eluent pH on the separations and found that acidic conditions created pH gradients due to protons liberated during metal–HIBA complexation. Also, between pH 3.3 and 4.5, pump and tubing components would begin dissolution and this interfered with the separation of the last seven metal ions—Gd(III) to Ce(III). The authors optimized the separation by adjusting the eluent pH above 4.5, and by increasing the buffer capacity by using higher total concentrations of HIBA. The best resolution was recorded using the 5-μm bonded ion-exchanger. A separation of the lanthanides using this column is reproduced in Fig. 6.5. The authors noted that eluents prepared with HIBA interfered with quantification based upon post-column formation of lanthanide–Alizarine Red S complexes. This kind

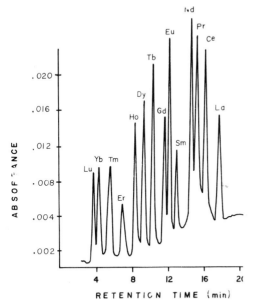

Fig. 6.5. Separation of the lanthanides. Column: 5-μm Nucleosil SCX. Eluent: linear program from 18–70 mM HIBA over a 20-min period, pH 4.6. Flow rate: 0.8 ml min^{-1}. Sample: 10 μl of a solution containing 10 μg ml^{-1} of each lanthanide. Detection: absorption at 600 nm after post-column reaction with Arsenazo I. (Reprinted with permission from ref. [3], copyright 1979 American Chemical Society.)

References pp. 165–166

of interference was not as pronounced if PAR or Arsenazo I was used as the post-column reagent. As a result, the optimal method for the analysis of lanthanide ions utilized post-column formation of either PAR, or Arsenazo I complexes.

In order to support research into the feasibility of deep geological disposal of used nuclear fuel, Kerr et al. [17] investigated HPLC methods to measure trace levels of uranium in solutions containing high concentrations of dissolved salts. Reverse-phase columns dynamically modified by the adsorption of ionic modifiers (surfactants) were used for these assays. The authors chose this method of separation due to the ease with which column specificity and capacity could be changed by varying the type and/or concentration of the mobile phase modifier.

The first step of the protocol developed consisted of passing a solution of HIBA through a reverse-phase enrichment cartridge in order to concentrate uranium from solution. This approach permitted the selective retention of uranium, while contaminants and possible interferents were flushed to waste. After a suitable period of time, the uranium was back-flushed from the enrichment cartridge and separated on a C_{18} analytical column using a solution containing 0.22 M HIBA and 25 mM pentanesulfonate, pH 3.5. In order to selectively monitor the uranium peak, the column effluent was mixed with Arsenazo III in a post-column reactor and the absorbance of the resulting complexes at 658 nm was monitored.

A high salt solution simulating the composition of natural waters found deep in granite rock formations was prepared and analyzed using this method, and the results showed the uranium peak to be well resolved from other peaks in the chromatogram. Also, the uranium signal was not affected by any potential interferent, and there was no appreciable deterioration in the chromatogram. The precision of the uranium quantification was evaluated by repeat analysis over several weeks using a control solution containing high concentrations of salt. At the 20 ng level, the precision of the method was 4% (2σ).

6.2.7. 8-Hydroxyquinoline (oxine)

Lakata et al. [18] investigated the suitability of HPLC methods for the trace speciation of a broad range of organotin compounds. The ability of both normal-phase and reverse-phase systems to resolve mixtures of mono-, di-, tri-, and tetraorganotin compounds was studied. Oxine was used in each mobile phase to aid the elution of the organotin species, and also to permit sensitive detection of each analyte using the absorbance

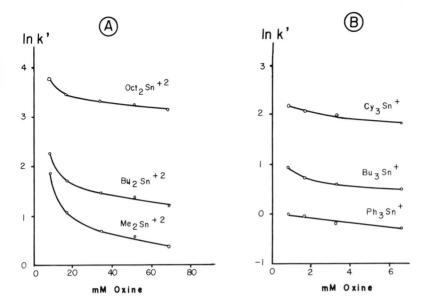

Fig. 6.6. Effect of oxine concentration on retention for (A) dialkyltin and (B) triorganotin compounds. Column: C_{18}, 150 mm × 3.2-mm i.d. Mobile phase: methanol–water–oxine (85 : 15 : variable, v/v/w); flow rate 1 ml min^{-1}. Bu = butyl, Me = methyl, Oct = octyl, Cy = tricyclohexyl, and Ph = phenyl. Detection: (A) photometric detection at 380 nm; (B) post-column photodegradation, followed by reaction with oxine, photometric detection at 380 nm. (Reproduced with permission from ref. [18], copyright 1984, Springer-Verlag.)

of those organotin–oxine complexes formed.

A study involving the retention of diorganotin compounds on a C_{18} reverse-phase column using mobile phases comprised of oxine in 85 : 15 (v/v) methanol–water revealed retention to be inversely dependent upon the oxine concentration in the mobile phase. A plot of capacity factor (k') as a function of oxine concentration for dimethyl-, dibutyl-, and dioctyltin is reproduced in Fig. 6.6A. Capacity factors for triorganotin species were less affected by changes in the oxine concentration than the diorganotin compounds (Fig. 6.6B). The authors also note that tetraorganotin retention was not influenced by different concentrations of oxine. Mono-organotin compounds were also studied, but these species required mobile phases with high percentages of water, and these conditions caused oxine to precipitate.

Based upon these results, the authors suggest three possible retention mechanisms, two of which are dependent upon the adsorption of oxine to the stationary phase. One mechanism assumes a mobile phase equilibrium between the "pure" organotin compounds and those which are bound by oxine. According to this mechanism, the position of equilibrium would determine the retention of the analyte, wherein the "pure" organotin compounds would experience weak complex formation interactions with the oxine adsorbed on the stationary phase, and the classic oxine complexes would be highly retained due to strong interaction with adsorbed oxine.

The second mechanism assumes that the stationary phase is dynamically coated with oxine. As a result, the coordination compounds are dynamically equilibrated between the mobile and stationary phases. The third mechanism proposed, and the one most favored by the authors, is based upon retention by hydrophobic interactions. The authors offer linear and positively correlated graphs of ln k' versus the number of carbon atoms in each organotin species in support of this mechanism.

6.2.8. Lactic acid

In a paper which emphasized the development of a post-column reaction for the inverse photometric determination of trace metals, Jones et al. [19] used strong-acid cation-exchange columns with a lactate eluent in order to separate the metal ions Cu(II), Fe(III), Ni(II), Zn(II), Co(II), Fe(II), Cd(II), Mn(II), and Mg(II). The authors indicate that one of their main concerns was to separate the metal ions in as short a time as possible, and as a result, used silica-based phases as these were more efficient than resin-based columns of similar particle size. In a previous publication from this group [23], eluents containing either citric acid or tartaric acid were used to separate the divalent ions of Cu, Zn, Ni, Co, and Cd on resin-based columns. When either citric acid or tartaric acid mobile phases were used with the silica phases, however, the lower exchange capacity of the silica-bonded columns caused the metals to elute too close to the solvent front to be of any value. As a result, the authors studied the effects of weaker chelating acids (acetic, formic, lactic, and succinic) to elute and separate the nine metal ions. The results using lactic acid showed metal ion retention to be highly dependent upon lactate concentration, enabling the separation to be optimized with small variations in the lactate concentration.

6.2.9. Phthalic acid

Jenke and Pagenkopf [20] reported on the occurrence of cationic interferences during the anion analysis of aqueous samples using nonsuppressed chromatography and conductivity detection. A 2.0 mM potassium hydrogen phthalate (KHP) eluent was used to separate some common anions on a Vydac silica-bonded anion-exchange separator column. Measurement of the collected eluent fractions by flame atomic absorption indicated coelution of Cl^- with Pb^{2+}, Br^- with Zn^{2+}, and Cu^{2+} with NO_3^-. Even though the elution order which was reported (Pb–Zn–Cu) for the metal ions was not a prime consideration of this study, it did prompt Siriraks et al. [21] to undertake a similar study four years later.

Armed with several anion-exchange columns and an eluent solution of 2.0 mM KHP (pH 5.0), Siriraks et al. set out to determine why the elution order they observed (Zn–Pb–Cu) was different than that reported by Jenke and Pagenkopf [20]. They studied the chromatographic behavior of Pb(II), Zn(II), and Cu(II) on four different anion-exchange columns in an attempt to elucidate the mechanism of retention for the three cations. The following columns were used in the study: silica-bonded from Vydac (302.IC); polystyrene-divinylbenzene (PS-DVB) from Hamilton (PRPX-100); polymethacrylate from Waters (IC-Pak); and a latex agglomerated PS-DVB from Dionex (AS-4).

The Dionex column caused each of the metal ions to be completely retained. The authors reasoned that this behavior was due to the presence of unreacted surface-sulfonated PS-DVB macroparticles which were capable of adsorbing the metal cations. A mixed retention mechanism was theorized for the other three columns. In addition to anion-exchange of negatively charged metal–phthalate complexes, the authors suggest that adsorption of neutral metal–phthalate complexes might contribute to the retention of each metal ion. This theory was based on the fact that since the stationary phase of the Hamilton, Waters, and Vydac columns were comprised of different materials, these materials would have different hydrophobicity which would lead to differences in adsorption of the neutral metal–phthalate complexes on each column. This was corroborated experimentally, as the metal ions were retained more strongly on the more hydrophobic column (PS-DVB from Hamilton). The dominant mechanism of retention was similar for each column because there were no differences in metal ion elution orders between the three columns.

6.2.10. Tartaric acid

Cassidy and Elchuk [22] investigated the potential of HPLC for the determination of trace amounts of metal ions in steel, nickel–

chromium–iron alloys, zirconium alloy, and uranium. The separation of Co(II), Cu(II), Fe(II), Mn(II), Ni(II), Pb(II), and Zn(II) was studied using a 13-μm Aminex A-5 PS-DVB cation-exchange resin, 100 mm $\times$ 4-mm i.d., and also a 10-μm Brownlee bonded-phase cation-exchange column. The seven cations could be separated on either column using pH gradients, but baselines were unstable using these conditions.

The authors found that Co(II) and Zn(II) could not be separated on the bonded-phase column. In addition, only low eluent concentrations could be used with the low capacity bonded-phase column, and this proved to be important when large concentrations of acid samples were injected. Separation of Co(II) and Ni(II), or Cu(II) and Ni(II), could not be achieved for samples with high levels of Ni(II) using the resin-based column and a citrate eluent. Tartrate eluents improved the separations, and the authors eventually used tartrate eluents for the assays. An important outcome of this study was the realization that either the gradient or the complexing agent could be changed in order to minimize interferences between the seven cations and the main constituents of those materials analyzed.

Cassidy and Elchuk [23] used PS-DVB and bonded C_{18} high efficiency reverse-phase columns to study separations for cationic and anionic species using anionic or cationic mobile phase modifiers, respectively. For the separation of metal cations, the modifiers used included C_{12} and C_{20} sulfates, and C_6 and C_8 sulfonates. Uniformly coated C_{20} sulfate columns were prepared by passing 500 ml of a 2.5×10^{-4} M solution of $C_{20}H_{41}SO_4^-$ in 3:1 (v) water/acetonitrile. After the coating procedure, the columns were used with aqueous mobile phases that did not contain any of the C_{20} sulfate. For the C_{12}, C_8, and C_6 modifiers, however, the hydrophobic portions were not large enough to prevent losses from the column to the eluent. If the lower molecular weight surfactants were to be used to prepare the exchange surface, it was necessary to include them in the mobile phase. For each of the separations developed in this study, tartrate was used in the mobile phase.

The authors indicated that aqueous solutions of the C_{12}, C_8, and C_6 modifiers equilibrated rapidly with the C_{18} column, and separations using these modifiers were similar to those obtained with C_{20} sulfate. The authors also noted that peak symmetry was generally better than previously recorded by researchers using bonded-phase ion-exchangers or PS-DVB ion-exchangers. The overall performance of this chromatographic technique is illustrated in Fig. 6.7 for the isocratic separation of the divalent metals Cu, Pb, Zn, Ni, Co, and Mn. The mobile phase used contained 45 mM tartrate, 10 mM hexanesulfonate, pH 3.4. These re-

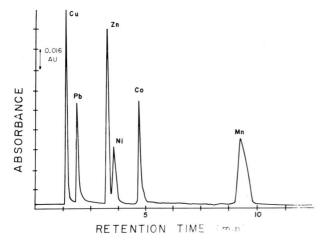

Fig. 6.7. Separation of six divalent metal ions. Column: Supelcosil 5-μm C_{18} silica-bonded phase. Eluent: 45 mM tartrate, 10 mM hexanesulfonate, pH 3.4. Flow rate: 1.5 ml min^{-1}. Sample: 20 μl of a solution containing 2.5–50 μg of each metal ion. Detection: absorption at 530 nm after post-column reaction with PAR. (Reprinted with permission from ref. [23], copyright 1982 American Chemical Society.)

sults indicate that metal tartrate equilibria do not limit the attainment of the rapid mass transfer required for high performance separations. In the case of manganese, however, a slow rate of attainment of Mn–tartrate equilibria might have caused the appearance of a small shoulder on the Mn peak.

Jones et al. [24] separated the metal ions Cu(II), Zn(II), Ni(II), Co(II), and Cd(II) on a sulfonated polystyrene resin using a tartrate eluent, and studied the potential for detecting these metal ions using a post-column reaction based on the chelating agent dithizone and inverse photometry. The chromatogram in Fig. 6.8 is reproduced from their work and shows that the five metal ions were separated under 30 min. Using sample volumes of 100 μl, the authors indicate the limits of detection to be between 1 and 8 ng for the metals. This small range is consistent with inverse photometric detection, since a decrease in the absorbance of the reagent is measured for each metal eluting from the column.

Sevenich and Fritz [2] used eluents containing the ethylenediammonium cation and either tartrate or α-hydroxyisobutyrate as the complexing anion to separate polyvalent metal cations on cation-exchange columns. The emphasis of this study was to achieve more selective chromatographic separation of metal ions, and also to show that the metal ions could be sensitively detected using conductimetry.

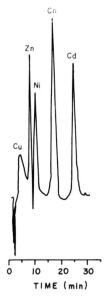

Fig. 6.8. Chromatogram of some divalent transition metals. Column: 11-μm Aminex A9 polystyrene resin, 90 mm × 5-mm i.d. Eluent: 0.2 M tartrate, pH 4.25. Detection: post-column reaction with dithizone, inverse photometry, 590 nm. Sample: 0.5 μg Cu(II), 1.0 μg Zn(II), 1.5 μg Ni(II), 1.5 μg Co(II), and 1.0 μg Cd(II). (Reproduced with permission from ref. [24], copyright 1983, Elsevier Science Publishers.)

The authors initially examined the separation of polyvalent metal cations on a 20-μm low capacity sulfonated gel-type resin using several different complexing agents (not listed), and found the best separations to be achieved using tartrate eluents. Those mobile phases which contained approximately equal molar concentrations of ethylenediammonium cation and tartrate anion produced well-formed peaks for divalent metal ions and also for some trivalent lanthanide cations.

In Fig. 6.9A is reproduced the chromatogram illustrating the separation of the divalent ions Zn, Co, Mn, Cd, Ca, Pb, and Sr, using an 1.5 mM ethylenediamine and 2.0 mM tartrate eluent, pH 4.0. Separation of the trivalent lanthanides Lu, Tm, Ho, Gd, Nd, Pr, Ce, and La (Fig. 6.9B) was achieved using a 2 mM ethylenediamine and 2.0 mM tartrate eluent, pH 4.5. Ethylenediamine was found to be an essential component of the mobile phase, as eluents containing ammonium tartrate were ineffective at eluting the metal cations. The authors assume elution to be based on a combined mass action "pushing" effect of the ethylenediammonium cation, and a complexing or "pulling" effect of the tartrate anion. Using

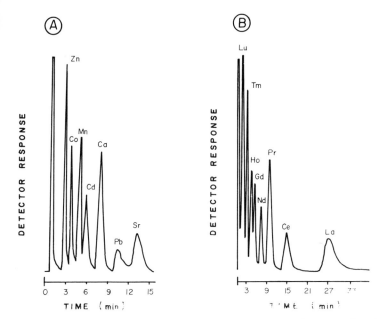

Fig. 6.9. Separation of some (A) divalent and (B) trivalent metal cations. Mobile phases: (A) 1.5 mM ethylenediamine and 2.0 mM tartrate, pH 4.0; (B) 2 mM ethylenediamine and 2.0 mM tartrate, pH 4.5. Column: 20-μm, 350 mm × 2-mm i.d., sulfonated cationexchange resin. Detection: conductivity. (Reprinted with permission from ref. [2], copyright 1983 American Chemical Society.)

an ion-exchange equilibrium as a basis, these investigators modified an ion-exchange retention equation to account for the fact that an injected sample of free metal ion will be bound by the complexing agent used in the mobile phase. The authors tested the validity of the equation by measuring the retention times for a number of cations using various mobile phase concentrations of ethylenediamine while holding steady the tartrate concentration and pH. The experimental data for the divalent ions correlated well with the results predicted using the equation developed, but the data recorded for the trivalent ions were not as promising.

Schmidt and Scott [25] developed a complete chromatographic system for the determination of some divalent metal cations. The system was very similar to the one used by Cassidy and Elchuk [23] in that a C_{18} bonded-phase column was used with an eluent containing 45 mM tartrate and 10 mM hexanesulfonate to separate the metal ions. Although there was a slight difference in the pH of the eluent used for the two studies (pH 3.1 and 3.4, resp.), the chromatogram reproduced

References pp. 165–166

144

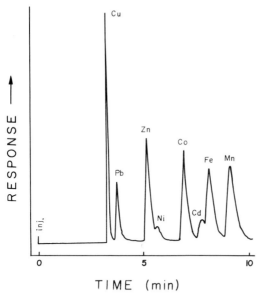

Fig. 6.10. Separation of eight divalent metal ions. Column: 10-μm C_{18} silica-bonded phase. Eluent: 45 mM tartrate, 10 mM hexanesulfonate, pH 3.1. Flow rate: 1.0 ml min^{-1}. Sample: 6 μl sample volume, approximately 1 μg of each metal. Detection: absorption after post-column reaction with PAR. (Reproduced from ref. [25], copyright 1984, The Royal Society of Chemistry.)

in Fig. 6.10 illustrates that the same elution order was observed for those metals common to both studies.

Schmidt and Scott's use of a post-column reaction of the metal ions with PAR to form complexes with high molar absorptivities was also adapted from Cassidy and Elchuk's paper. However, Schmidt and Scott built their own small volume absorption detector using a light emitting diode (LED) as the light source and a photocell as the sensing device. Another novel aspect of this paper was the fact that instead of using an additional column before the analytical column to achieve sample concentration, it was shown that metal ions could be efficiently concentrated on the analytical column by injecting the sample into an appropriate mobile phase system, and that the elution process could be initiated by switching to the mobile phase eluent. The authors indicate that metal ion retention is based upon ion-exchange with the sulfonate groups of hexanesulfonates which have sorbed onto the stationary phase, and that retention decreased with an increase in the concentration of tartrate in the mobile phase.

In order to concentrate metal ions, the authors injected milliliter volumes of metal ions into a phase system void of tartrate, and then changed to the hexanesulfonate/tartrate containing mobile phase to elute the metal ions. Peak height measurements were shown to be linearly related to injection volume in a study covering the range from 0.25 to 2.0 ml. Chromatograms were shown for the concentration of a 1 ml sample of 1 ppm Cu(II), and also 2.3 ml samples of 50 ppb Cu(II), Zn(II), Co(II), and Mn(II). In all cases, the peaks were very symmetric. The authors also note that this system worked well for the determination of cations present at the ppb level in samples containing high concentrations of salt (seawater) and/or organic materials (urine).

A fundamental study involving the micelle exclusion chromatographic separation of Fe(III), Cu(II), Pb(II), Zn(II), Ni(II), Co(II), Fe(II), and Mn(II) was described by Okada [26].The metals were separated on a 5-μm, Asahipak GS-310H poly(vinyl alcohol) gel-type size exclusion column, 250 mm × 7.6-mm i.d., and sodium dodecyl sulfate (SDS) was used to form the micellar eluent. Because the stationary phase was effective at excluding compounds with molecular weights greater than 4×10^4, the micelles of SDS—with an aggregation number of 62—were only partially excluded under these conditions. The complexing ligands citrate and tartrate were used to elute the metal ions, and Fig. 6.11 illustrates the retention of each metal ion as a function of the mobile phase concentration of tartrate (Fig. 6.11A) and citrate (Fig. 6.11B). The relative retentions for most of the metals were similar for tartrate and citrate mobile phases, but some exceptions were evident, i.e., Zn(II), Pb(II), and Ni(II).

Okada recognized that three different mechanisms might contribute to the retention of a metal species: (1) the partitioning of neutral metal complexes into the hydrophobic interior of the micelle core; (2) partitioning based upon the electrostatic attraction of free metal ions to the charged groups on the periphery of the micelle; and (3) ion-pairing or ion-interaction effects in the interior part of the stationary phase called the inner solution, due to the permeation of monomeric surfactants within that region. The author based the derivation of an equation to calculate metal ion retention on the electrostatic attraction model, and used the equation to calculate the retention of Co(II), Zn(II), and Pb(II) for different mobile phase concentrations of SDS and tartaric acid. Good correlation between calculated and experimentally measured retention times indicated the equation was valid for those mobile phases. The same equation was used for mobile phases containing SDS + citrate,

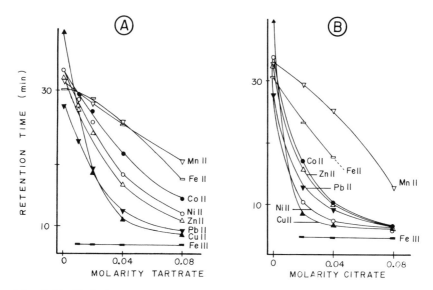

Fig. 6.11. Variation of metal ion retention time as a function of the mobile phase concentration of (A) tartaric acid, and (B) citric acid. Column: 5-μm Asahipak GS-310H poly(vinyl alcohol) gel-type size exclusion, 250 mm × 7.6-mm i.d.. Mobile phases: (A) tartaric acid + 0.05 M SDS, pH 4.4; (B) citric acid + 0.05 M SDS, pH 4.3. Detection: post-column reaction with PAR, photometric detection at 540 nm. (Reprinted with permission from ref. [26], copyright 1988 American Chemical Society.)

but the retention times calculated for the three metals did not correlate with experimental values. The author attributed the poor correlation to the fact that those metals studied formed mixtures of neutral and anionic metal complexes, and that the equation was derived without consideration of the influence of these species.

It is interesting to note that the elution order reported by Okada for the micellar exclusion separation of Cu(II), Pb(II), Zn(II), Ni(II), Co(II), and Mn(II) was the same as the order reported by Cassidy and Elchuk [23], and Schmidt and Scott [25], for reverse-phase separation of these ions. The latter two studies both utilized C_{18} columns and eluents containing 45 mM tartrate and 10 mM hexanesulfonate. The surprising aspect of these three studies is the fact that similar elution orders could be achieved using size exclusion columns with a micellar mobile phase, as with reverse-phase columns and non-micellar mobile phases.

6.2.11. Mixed complexing eluents

6.2.11.1. *Citric acid and oxalic acid*

Boyle and co-workers [27] undertook a study to determine whether an HPLC method based upon luminol chemiluminescence detection could be used for the analysis of Co(II) in seawater and freshwater samples at the 10^{-10} mol kg^{-1} level. Their preliminary results showed weak chemiluminescent signals for those chromatographic systems in which cobalt eluted from the column in a complexed form: chemiluminescence from the luminol reaction is dependent upon catalysis by free metal ions. It was also observed that the chemiluminescent signal was not severely depressed when moderately strong complexing agents were used in the mobile phase. These results ultimately led the researchers to use weakly complexing eluents in order to elute cobalt in a relatively uncomplexed form, and thus ensure maximum sensitivity from the chemiluminescence reaction.

The authors studied the chromatographic behavior of Co(II), V(II), Cu(II), and Fe(II) using a Dionex HPIC-CS2 cation-exchange column and an eluent containing oxalic acid and citric acid. It was found that the retention of the metals could be controlled by varying the eluent pH, or the mobile phase concentration of either complexing agent. An increase in pH resulted in lower k' values for the metals. This effect was ascribed to a shift in the metal–ligand equilibrium due to an increase in ionized ligand concentration with pH. The authors also observed a decrease in k' with increases in the mobile phase concentration of either citric acid or oxalic acid. The separation of a standard containing V(II), Cu(II), Co(II), and Fe(II) prepared in 0.1 M HCl is reproduced in Fig. 6.12. The detection limits for these metals were 0.15, 0.3, 0.04, and 0.3 nM, respectively. Seawater samples were also analyzed using this system, but because the background salts proved to be a more powerful eluent than either tartrate or citrate, Co(II) eluted in the solvent front. In the final method for seawater samples, cobalt was separated from the matrix by a double extraction using ammonium pyrrolidinedithiocarbamate (APDC) and CCl_4.

6.2.11.2. *Citric acid and tartaric acid*

The group at Dionex reported on new technology and illustrated separations for alkali and alkaline earth metals, transition metals, and amines [28]. An efficient separation of divalent Ni, Zn, Co, Pb, Fe, Cd,

148

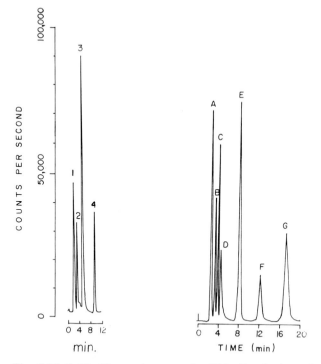

Fig. 6.12 (left). Chromatogram of a 500 μL solution of metal ions prepared in 0.1 M HCl. Column: Dionex cation-exchange HPIC-CS2. Mobile phase: 0.1 M oxalic acid and 75 mM citric acid, pH 4.1 using LiOH; flow programming was employed, 0.5 ml min^{-1} for the first 5 min and 2.0 ml min^{-1} for the next 7 min. Peaks: *1* = V(II), (74 nM); *2* = Cu(II), (64 nM); *3* = Co(II), (26 nM); *4* = Fe(II), (110 nM). Detection: chemiluminescence post-column reaction using luminol, reagent pH = 10.6. (Reprinted with permission from ref. [27], copyright 1987 American Chemical Society.)

Fig. 6.13 (right). Cation-exchange separation of transition metals. Column: HPIC-CS2. Mobile phase: 12 mM citric acid, 40 mM tartaric acid, pH 4.1 with LiOH; flow rate 1.0 ml min^{-1}. Peaks: *A* = Ni(II), (1 ppm); *B* = Zn(II), (0.5 ppm); *C* = Co(II), (0.5 ppm); *D* = Pb(II), (5 ppm); *E* = Fe(II), (3 ppm); *F* = Cd(II), (3 ppm); *G* = Mn(II), (2 ppm). Detection: post-column reaction using PAR, photometric detection at 520 nm. (Reproduced from ref. [28].)

and Mn was achieved on a cation-exchange column using an eluent containing tartaric acid and citric acid (Fig. 6.13).

6.2.11.3. *Diglycolic acid and oxalic acid*

Separations of lanthanide ions are difficult to achieve by classic ion-exchange techniques due to the general similarities in electronic

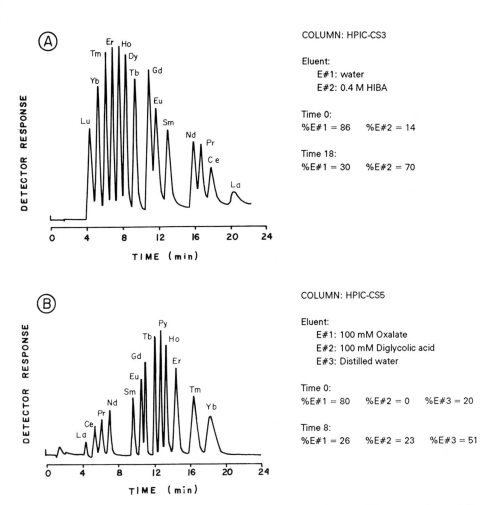

Fig. 6.14. Gradient cation-exchange (A) and anion-exchange (B) separation of the lanthanides. Detection: post-column reaction using PAR, photometric detection at 520 nm, 0.5 AUFS. Sample: 10 ppm each metal, 50 μl sample volume. Note: Yb(III) and Lu(III) co-eluted for the anion-exchange separation. (Reproduced with permission from ref. [29].)

structure and the state of hydration of these ions. However, by selecting a complexing agent which has small incremental differences in the magnitude of formation constants for successive lanthanides in the series—yet a large range across the series—Heberling et al. [29] were able to separate the lanthanides by either cation-exchange or anion-exchange chromatography. The chromatogram in Fig. 6.14A was recorded using a cation-exchange column and a gradient of 0.4 M α-HIBA; here the ions elute in order of decreasing atomic number. By changing to an anion-exchange column and eluents of 100 mM oxalate and 100 mM diglycolic acid, these researchers showed a reversed order of elution for the lanthanide ions (Fig. 6.14B). The authors were able to separate the lanthanides by cation-exchange because complexation of these metals with α-HIBA leads to the formation of both neutral and cationic lanthanide–HIBA complexes, and each metal retains some cationic character at equilibrium. On the other hand, complexation with oxalate and diglycolic acid results in the formation of predominately anionic lanthanide complexes which are amenable to anion-exchange chromatography. These chromatograms are illustrative of the efficient separations that can be achieved for inorganic species using *in situ* complexation reactions, and they also exemplify the high selectivity that is possible when the secondary chemical equilibria reactions are firmly established and under the control of the analyst.

6.2.11.4. α-Hydroxyisobutyric acid (HIBA) and tartaric acid

Karcher and Krull [30] used complexometric calculations to determine the mobile phase concentrations of HIBA and tartaric acid needed to fine tune their separation of eleven metal cations on C_8 and C_{18} silica-based reverse-phase columns which had been dynamically modified with n-octanesulfonate. Isocratic elution was used to separate the metals into three distinct windows; each window corresponding to one of the three different valence states spanned by the eleven metal cations. The metals eluted in the order of increasing valence, with the sole exception of La(III) which did not elute within the trivalent ion window. Figure 6.15 illustrates the separation of Zr(IV), Ga(III), Sc(III), Y(III), Al(III), In(III), Zn(II), La(III), Cd(II), Ca(II), and Mg(II) using a C_{18} column and an eluent comprised of 2.27 mM n-octanesulfonate, 8.18 mM tartaric acid, 52.9 mM HIBA, and 10.7% (v/v) methanol.

The authors indicate that the elution orders for samples of metal ions can be accurately predicted using complexometric calculations. The algorithm used requires inputs of mobile phase ligand concentrations, mobile phase pH, and the formation constants for reactions between metal ions and the mobile phase ligands. In addition to their own experimental results, the authors have successfully calculated elution orders for many of the literature reported *in situ* complexation-based separations which have been achieved on either cation-exchange or dynamically modified reverse-phase columns [31].

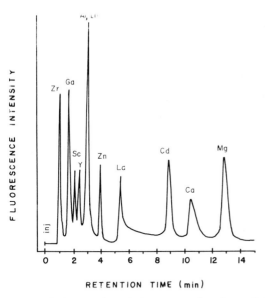

Fig. 6.15. Separation of eleven metal cations. Column: 5μm C₁₈ Lichrospher, 125 mm × 4.0-mm i.d. Eluent: 2.27 mM *n*-octanesulfonate, 8.18 mM tartaric acid, 52.9 mM HIBA, pH 4.1, and 10.7% (v/v) methanol. Sample: 20 μl loop volume; 23.4 ppm Zr, 8.4 ppm Ga, 5.4 ppm Sc, 3.1 ppm Y, 3.1 ppm Al, 10.5 ppm Zn, 5.5 ppm La, 23.1 ppm Cd, 81.4 ppm Ca, and 4.8 ppm Mg. Detection: post-column reaction with oxine, fluorescence measured using a 360 nm bandpass filter for excitation and 455 nm longpass filter for emission. (Reproduced from the *Journal of Chromatographic Science* by permission of Preston Publications, a Division of Preston Industries, Inc.)

6.3. ION-EXCHANGE SEPARATIONS

6.3.1. Complex equilibria

The retention of an eluite in liquid chromatography is based upon the distribution of the eluite between the stationary and mobile phase (primary equilibrium). By convention, any other equilibria that takes place in the mobile phase, or stationary phase, or both, are considered "secondary". In the past, manipulation of the mobile–stationary phase equilibrium distribution of the solute by using secondary chemical equilibria (SCE) was widely utilized in order to overcome low column efficiencies [32]. In spite of the fact that we now have columns with higher inherent efficiencies, SCE is still a widely practiced technique. The wide range of different chemistries that can be used to alter the mobile–stationary phase equilibrium to achieve better resolution or

References pp. 165–166

selectivity makes SCE a sensible alternative to the more conventional HPLC modes.

One example of SCE is the use of complexing eluents for the separation of metal species. This method generally involves using one or more complexing agents in the mobile phase in order to resolve a specific few, or a group, of metal ions. Generally, the chromatographic resolution in these systems can be controlled by adjustment of the mobile phase concentration of the complexing agent(s). Other attractive features of this technique include the potential to generate high column efficiencies, and an improvement in detection properties of metal complexes over the free ions.

A dilute solution of a weakly complexing ligand is the most frequently used mobile phase for this form of SCE, and these conditions generally favor the formation of one or more metal ligand species for each metal cation in the sample solution. The influence of these metal species on the liquid chromatographic retention of a metal cation is discussed in the following subsection. Some fundamental principles of complex equilibria are first reviewed in order to familiarize the reader with the types of reactions that may occur, along with the products which are likely to form, as a result of SCE reactions.

In a solution containing aquated metal ions M, and unidentate ligands L, there is a tendency for metal ions and metal complexes to engage in reactions which result in the addition or replacement of one or more of the ligands in the coordination spheres of the metal ions. The following equations and equilibrium constants can be used to describe such a system at equilibrium:

$$M + L \rightleftharpoons ML \qquad K_1 = \frac{[ML]}{[M][L]} \tag{2}$$

$$ML + L \rightleftharpoons ML_2 \qquad K_2 = \frac{[ML_2]}{[ML][L]} \tag{3}$$

$$\vdots \qquad\qquad \vdots$$

$$ML_{n-1} + L \rightleftharpoons ML_n \qquad K_n = \frac{[ML_n]}{[ML_{n-1}][L]} \tag{4}$$

There will be n such equilibria, where n represents the maximum coordination number of the metal. By convention, we have written the metal ion without specifying charge or degree of solvation. The K_i's are called stepwise formation constants (or stepwise stability constants).

The magnitude of these constants indicate the thermodynamic tendency of a species to exist (to be "stable") once equilibrium has been established in the system. Formation constants are not rate constants, and thus do not give insight as to how quickly equilibrium is established, or how rapidly the metal species interconvert in solution, vide infra.

It is equally appropriate to represent the thermodynamics of complex equilibria by using overall formation constants. These constants take the form β_i, the subscript i designating the total number of ligands added to the aquated free metal ion. The equations and overall equilibrium constants which describe the formation of complexes ML_1, ML_2, and ML_n are:

$$M + L \rightleftharpoons ML \qquad \beta_1 = \frac{[ML]}{[M][L]} = K_1 \qquad (5)$$

$$M + 2L \rightleftharpoons ML_2 \qquad \beta_2 = \frac{[ML_2]}{[M][L]^2} = K_1 K_2 \qquad (6)$$

$$\vdots \qquad\qquad \vdots$$

$$M + nL \rightleftharpoons ML_n \qquad \beta_n = \frac{[ML_n]}{[M][L]^n} = K_1 K_2 \ldots K_n \qquad (7)$$

The relationship between overall and stepwise formation constants is given by the following general expression:

$$\beta_j = K_1 K_2 K_3 \ldots K_j = \prod_{i=1}^{i=j} K_i \qquad (8)$$

If the pH and concentration of each ligand in solution are known, then these reactions and constants can be used to determine the equilibrium concentration of each metal species. In those instances in which the solution is also a liquid chromatographic eluent, such qualitative and quantitative information relates to the different types of eluite forms that are present in the mobile phase—free and complexed metal. This information is significant because the retention that is observed for a particular eluite is dependent upon the relative proportion of each different eluite forms present in solution. Those readers having utilized one of the more well known modes of secondary chemical equilibria (i.e., acid-base, ion-pairing, solute-micelle) may be well versed in how changes in the composition of the mobile phase can influence the relative proportions of eluite forms, and ultimately, the overall appearance of the chromatogram [33–35].

References pp. 165–166

6.3.2. Fundamentals of retention

Most of the papers reviewed in this chapter share a common theme in that the HPLC separations described are based on an ion-exchange type mechanism. However, a great disparity also exists among these papers: some separations were developed on strong-acid ion-exchange columns, whereas other separations were achieved using reverse-phase columns with eluents containing ion-interaction reagents. Notwithstanding the differences in equipment and mobile phases used for these forms of ion-exchange, a completely general equilibrium can be written to describe retention of a free metal cation M^{m+} on an ion-exchange surface:

$$m\,CS + M^{m+} \rightleftharpoons (M^{m+}) \cdot m\,S^- + m\,C^{1+} \tag{9}$$

In this equilibrium reaction, S represents a surface exchange group of $1-$ charge, M is a metal ion of $m-$ charge, and C^{1+} denotes a singly charged counterion. Those terms containing "S" indicate species that are retained by the exchange surface, whereas all other species are present in the mobile phase.

The equilibrium constant for this reaction is called the selectivity coefficient ($K_{C+}^{M^{m+}}$). The magnitude of the selectivity coefficient is an indication of the exchanger's affinity for a metal ion relative to a counterion. The selectivity coefficient for the above exchange equilibrium is given by:

$$K_{C+}^{M^{m+}} = \frac{[(M^{m+}) \cdot m\,S^-]\,[C^{1+}]^m}{[CS]^m\,[M^{m+}]} \tag{10}$$

Substituting the exchange capacity (Cap.) of the column for the stationary phase concentration of the counterion [CS], and k' for the ratio of the metal concentration in the stationary phase to the metal concentration in mobile phase $[(M^{m+}) \cdot m\,S^-]:[M^{m+}]$, an expression is obtained which relates retention of the free metal ion in terms of two known or easily measured parameters ($K_{C+}^{M^{m+}}$ and column capacity), and an experimentally controlled variable ($[C^{1+}]$):

$$\log k'_{M^{m+}} = \log K_{C+}^{M^{m+}} + m \log (Cap.) - m \log[C^{1+}] \tag{11}$$

Equation 11 can be used for the determination of capacity factors in ion-exchange systems, but only if the eluent is unreactive to the metal eluite cations. This requisite mandates that no strong metal–ligand complexes be formed within the column, and that each metal exist predominately as the free ion M^{m+} throughout the elution process.

Accordingly, eqn. 11 cannot be applied to ion-exchange separations which utilize complexing eluents.

A modified version of eqn. 11 was used by Sevenich and Fritz [2] to describe the retention of metal cations eluted from strong-acid exchange columns using complexing eluents. These authors added the term log $\phi_{M^{m+}}$ to account for the reduction of free metal ion concentration due to complexation with the tartrate used in the mobile phases. $\phi_{M^{m+}}$ represents the mole fraction of free metal ion present in solution, and is a quantity which can be calculated using complexometric methods. An equation to describe the above ion-exchange equilibria can be written using a format similar to that used by Sevenich and Fritz:

$$\log k'_{M^{m+}} = \log K^{M^{m+}}_{C^+} + m \log (\text{Cap.}) - m \log[C^{1+}] \, \phi_{M^{m+}} \tag{12}$$

This equation was obtained by substituting $[M'] \cdot \phi_{M^{m+}}$ for $[M^{m+}]$ in eqn. 10, where $[M']$ represents the total concentration of metal (free and complexed) in solution. The capacity factor ($k'_{M^{m+}}$) was then substituted for the ratio $[M^{m+} \cdot mS^-] : [M']$. Sevenich and Fritz tested the validity of this equation and found fairly good agreement for those divalent ions studied. Some trivalent metals were also tested, but these did not show the same degree of correlation as the divalent metals.

Sevenich and Fritz realized the retention measured for a metal to be dependent upon the fraction of free cation in solution. However, the retention of a metal is also dependent upon positive charged complexed forms of the metal. For instance, if a trivalent metal ion (M^{3+}) were to react with a singly charged ligand (L^{1-}) to form the metal species ML_1^{2+}, ML_2^{1+}, and ML_3 within the column, the various charges would cause each of these species to be retained differently by the exchange surface. The two ion-exchange equilibria that can be written for the cationic metal complexes are:

$$2\,CS + ML_1^{2+} \rightleftharpoons (ML_1^{2+}) \cdot 2S^- + 2\,C^{1+} \tag{13}$$

$$1\,CS + ML_2^{1+} \rightleftharpoons (ML_2^{1+}) \cdot 1S^- + 1\,C^{1+} \tag{14}$$

The singly charged counterions and surface exchange groups are represented by C and S, respectively. The selectivity coefficients for these equilibria are given by:

$$K^{ML_1^{2+}}_{C^+} = \frac{[(ML_1^{2+}) \cdot 2S^-][C^{1+}]^2}{[CS]^2 [ML_1^{2+}]} \tag{15}$$

$$K^{ML_2^{1+}}_{C^+} = \frac{[(ML_2^{1+}) \cdot 1S^-][C^{1+}]^1}{[CS]^1 [ML_2^{1+}]} \tag{16}$$

References pp. 165–166

By proceeding as outlined above, equations which describe the retention of each cationic complex can be derived.

$$\log k'_{ML_1^{2+}} = \log K_{C+}^{ML_1^{2+}} + (2) \log (\text{Cap.}) - (2) \log[C^{1+}] \, \phi_{ML_1^{2+}} \qquad (17)$$

$$\log k'_{ML_2^{1+}} = \log K_{C+}^{ML_2^{1+}} + (1) \log (\text{Cap.}) - (1) \log[C^{1+}] \, \phi_{ML_2^{1+}} \qquad (18)$$

Because there are three separate retention equations (eqns. 12, 17, 18) to account for the three cationic metal species in solution, one might expect to find three peaks in the final chromatogram. In reality, however, only one peak is usually evident. The number of peaks that appear is a function of the kinetics of the equilibrium processes occurring in solution. One peak is seen if the rate of all reversible and/or irreversible chemical equilibria associated with an eluite as it migrates through the column is fast relative to the elution time of the eluite. If interconversion between the metal–ligand species is slow, however, asymmetric or multiple peaks may result.

6.3.3. Retention dependence on weighted average charge

Foley [33] has indicated that for any eluite which can exist in more than one chemical form due to secondary chemical equilibria, retention is given by the summation of the weighted average capacity factors for each eluite form. The mole fraction of each of the eluite species is calculated and used as weighting factors in this treatment. For the trivalent metal ion system described above, M^{3+}, ML_1^{2+}, ML_2^{1+}, and ML_3 coexist in solution, and the capacity factor for the single peak observed is given by:

$$k'_{obs} = k'_{M^{3+}} \cdot \phi_{M^{3+}} + k'_{ML_1^{2+}} \cdot \phi_{ML_1^{2+}} + k'_{ML_2^{1+}} \cdot \phi_{ML_2^{1+}} + k'_{ML_3} \cdot \phi_{ML_3} \qquad (19)$$

This equation is concise, but it is often difficult or impossible to isolate the metal species of a dynamic system and preserve their integrity throughout the duration of an experiment designed to determine individual k'_i values. In our own studies, we have examined alternate forms of the retention equation and have found good correlation between predicted and experimental retention times by substituting the charge (Z_i) of each metal species for the k'_i terms in eqn. 19. Rewritten in this form, the relative retention for the trivalent metal M^{3+} is:

$$k'_{obs} \propto Z_{M^{3+}} \cdot \phi_{M^{3+}} + Z_{ML_1^{2+}} \cdot \phi_{ML_1^{2+}} + Z_{ML_2^{1+}} \cdot \phi_{ML_2^{1+}} + Z_{ML_3} \cdot \phi_{ML_3} \qquad (20$$

The k'_{obs} values calculated using this format are essentially weighted average charges. For samples containing several different metal cations, our experiments show that weighted average charges can be used to predict the elution order of the cations, and also to aid in the optimization of chromatographic resolution [31]. In the former application, the composition of the mobile phase is known, and equations similar in form to eqn. 20 are developed for each metal cation in the sample. The equilibrium distribution of metal species—in terms of mole percentages—is then calculated for each of the metal cations in the sample using the following data: the pH of the mobile phase, concentrations of all ligands in the mobile phase, and the formation constants which describe equilibrium reactions between the metal eluite cations and the mobile phase ligands. Once the mole percent terms have been calculated, each is multiplied by the charge of the appropriate metal species, and all terms are summed to obtain k'_{obs} values for each metal in the sample. Ranking the metals in increasing value of k'_{obs} reflects the order of elution expected for a cation-exchange separation. If the metals are separated via an anion-exchange mechanism (some metals, for example, form negative charged EDTA or chloro complexes), then the opposite is true, and the metal experiencing the greatest retention will be the one with the smallest k'_{obs}.

For those studies involving optimization of chromatographic resolution, a range of ligand concencentrations or mobile phase pH is selected, and a computer is used to calculate ϕ_i for each metal species at fractional increments across the range of pH or ligand concentration. Multiplication of each term by the appropriate metal species charge permits the determination of k'_{obs} for each metal in the sample. The analyst then selects and experimentally verifies those mobile phase conditions which seem to best satisfy the chromatographic efficiency and resolution requirements for the assay. As might be expected [36], the agreement between observed and calculated elution orders is highly dependent upon the availability and accuracy of the formation constants used for these calculations. In the next section, the procedure used in the calculation of a weighted average charge is outlined. In addition, several examples are presented that illustrate the correlation between experimentally measured retention, and the retention expected based upon the calculated weighted average charge.

References pp. 165–166

6.3.4. Weighted average charge calculations

The first step in the calculation of a weighted average charge is to identify and quantify each of the metal species in solution. Equations to represent the total concentration—free ion and all complexed forms— are written for each metal in the sample solution:

$$C_M = [M] + [ML] + [ML_2] + \ldots + [ML_n] \tag{21}$$

The concentration of each metal complex is then rewritten using equations similar in form to eqns. 5–7:

$$C_M = [M] + \beta_1 [M][L] + \beta_2 [M][L]^2 + \ldots + \beta_n [M][L]^n \tag{22}$$

The mole fraction of each metal species is then calculated and used to quantify the species in solution. The mole fraction of a mononuclear metal complex is represented by:

$$\phi_i = \frac{[ML_i]}{C_M} = \frac{\beta_i [M][L]^i}{[M] + \beta_1 [M][L] + \beta_2 [M][L]^2 + \ldots + \beta_n [M][L]^n} \tag{23}$$

Equation 23 readily simplifies to show that the mole fraction of any mononuclear complex ML_i is independent of the concentration of metal ion in solution:

$$\phi_i = \frac{\beta_i [L]^i}{1 + \beta_1 [L] + \beta_2 [L]^2 + \ldots + \beta_n [L]^n} \tag{24}$$

The concentration of free metal ion is given by:

$$\phi_{M^{m+}} = \frac{[M^{m+}]}{C_M} = \frac{1}{1 + \beta_1 [L] + \beta_2 [L]^2 + \ldots + \beta_n [L]^n} \tag{25}$$

The significance of eqns. 24 and 25 is that the mole fraction of all metal species in solution can be calculated if the following information is known: the pH of the solution, the concentration(s) of all ligand(s) in solution, and the formation constants for all metal–ligand equilibria that may occur. In the event that several different ligands are present in solution, the mole fraction equations for each metal must account for all complexed forms of the metal. For example, if L and Q are used to denote two different ligands, and β_{i_L} and β_{i_Q} represent overall formation constants for metal complexes of each ligand, respectively, then the mole fraction of MQ_i could be represented using the following general equation:

$$\phi_{i_Q} = \frac{\beta_{i_Q}[Q]^i}{1 + \beta_{1_L}[L] + \beta_{2_L}[L]^2 + \ldots \beta_{1_Q}[Q] + \beta_{2_Q}[Q]^2 + \ldots} \tag{26}$$

One example calculation is shown for an early ion-exchange separation [2] to illustrate the correlation between the observed retention and the relative retention predicted based upon the above equations. Table 6.1 lists the equilibria considered, the value of the corresponding equilibrium constants, the calculated mole percent of each metal species present in solution, and the weighted average charges. The ionic strength of the mobile phase affects the value of activity coefficients for charged species in solution, and it is therefore necessary to know the exact composition of the mobile phase in order to account for these influences (i.e., concentration constants are affected by ionic strength). The Davies equation [37] is recommended in order to improve the agreement of the calculated activity coefficients with the observed values.

The chromatogram illustrating the separation of these seven cations is reproduced in Fig. 6.16. The calculated elution order, based upon the weighted average charge algorithm, is the same as the elution order observed experimentally. This good agreement is attributed to the similar chemical and physical properties between lanthanide ions, and also to the fact that formation constants for the metal–HIBA equilibria were taken from the same literature and experimental source. The corroboration between calculated and experimental elution orders for samples comprised of cations with dissimilar chemical and physical properties, however, is usually not as good.

In our own investigations, a computer programmed with the weighted average charge algorithm was used to determine the concentration(s) of complexing agent(s) to use in the mobile phase in order to achieve optimum separations for samples of metal cations. An isocratic HPLC method selective for eleven cations was developed using this technique. The eleven cation sample contained four divalent ions, six trivalent ions, and one tetravalent ion. Selectivity of the method was imparted by fluorescence detection, as each cation formed fluorescent metaloxinates when mixed on-line with a reagent solution of oxine. The equilibria considered for these metals are listed in Table 6.2, along the values of the formation constants, the equilibrium distribution calculated for each metal species, and also the weighted average charge calculated for each cation. In this table, metal–HIBA equilibria are represented by ML_i terms, and metal–tartrate equilibria by MT_i.

Although there are no literature reported α-HIBA equilibria for aluminum, indium, or scandium, experimental evidence showed the elution

TABLE 6.1

CALCULATED WEIGHTED AVERAGE CHARGE FOR LANTHANIDE IONS IN A
SOLUTION CONTAINING 3.0 mM α-HIBA AT pH 4.5

Metal ion	Equilibrium	Log β	Mole percent	Weighted average charge
Tb(III)	M	–	5.41	+138.21
	ML	3.11	36.70	
	ML$_2$	5.63	48.58	
	ML$_3$	7.43	9.31	
Dy(III)	M	–	3.10	+121.89
	ML	3.27	30.39	
	ML$_2$	5.90	51.82	
	ML$_3$	7.87	14.69	
Ho(III)	M	–	2.66	+118.36
	ML	3.31	28.53	
	ML$_2$	5.98	53.34	
	ML$_3$	7.96	15.47	
Er(III)	M	–	2.25	+111.60
	ML	3.35	26.49	
	ML$_2$	6.04	51.87	
	ML$_3$	8.13	19.39	
Tm(III)	M	–	1.45	+105.42
	ML	3.52	25.24	
	ML$_2$	6.22	50.58	
	ML$_3$	8.39	22.73	
Yb(III)	M	–	0.90	+94.28
	ML	3.64	20.77	
	ML$_2$	6.42	50.03	
	ML$_3$	8.69	28.30	
Lu(III)	M	–	0.77	+87.98
	ML	3.67	18.94	
	ML$_2$	6.47	47.79	
	ML$_3$	8.82	32.50	

Note: L = hydroxyisobutyrate ion, M = lanthanide ion. β are overall formation constants. Formation constant data from ref. [38].

of these metals to be a function of the concentration of α-HIBA used in the mobile phase. The fact that elution of these cations was not influenced by moderate changes in the mobile phase concentrations of non-complexing ligands implicated that these cations were indeed complexed by α-HIBA. Complex equilibria for lactic acid (2-hydroxypropanoic

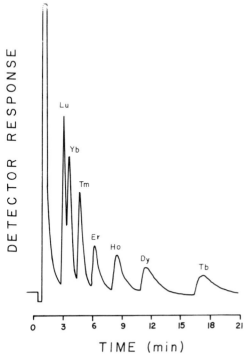

Fig. 6.16. Ion-exchange separation of seven lanthanide cations. Mobile phase: 4.0 mM ethylenediamine and 3.0 mM α-HIBA, pH 4.5. Column: 20-μm, 35 cm × 2-mm i.d., sulfonated cation exchange resin. Detection: conductivity. (Reprinted with permission from ref. [2], copyright 1983 American Chemical Society.)

acid)—structurally similar to HIBA—have been used in Table 6.2 to approximate the equilibrium reactions between α-HIBA and Al(III), In(III), and Sc(III). The predicted elution order—obtained by ordering the metals in increasing magnitude of the calculated weighted average charge—is listed below, along with the experimentally observed order recorded for a mobile phase comprised of 58.2 mM α-HIBA and 8.8 mM tartaric acid, pH 4.1.

observed: Zr Ga Sc Y Al In Zn La Cd Ca Mg
predicted: Zr Ga Sc Al In Y La Zn Cd Ca Mg

The agreement between the observed (see Fig. 6.15) and predicted elution order is very good, but two discrepancies do exist: (Y and Al-In) and (Zn and La). These inconsistencies arise because the relative retention equation (eqn. 20) does not rigorously account for the retention of

TABLE 6.2

CALCULATED WEIGHTED AVERAGE CHARGE FOR SOME METAL CATIONS IN A SOLUTION CONTAINING 58.2 mM α-HIBA AND 8.8 mM TARTARIC ACID, pH 4.1

Metal ion	Equilibrium	Log β	(IS)	Ref.	Mole percent	Weighted average charge
Zr(IV)	M	–		–	0.00	−100.00
	MT(OH)3	43.10	(0.1)	39	100.00	
Ga(III)	M	–		–	0.00	−99.43
	MT	4.93	(0.1)	39	0.28	
	MT2	9.76	(0.1)	39	99.72	
Sc(III)	M	–		–	0.00	−99.18
	ML	3.78	(1.0)	39	0.00	
	ML2	8.02	(1.0)	39	0.41	
	MT2	12.50	(0.1)	40	99.59	
Al(III)	M	–		–	0.00	−99.07
	ML	0.85	(0.1)	39	0.00	
	ML2	2.92	(0.1)	39	0.00	
	ML3	3.79	(0.1)	39	0.00	
	MT	5.32	(1.0)	40	0.46	
	MT2	9.77	(1.0)	40	99.54	
In(III)	M	–		–	0.90	−90.17
	ML	3.14	(0.2)	39	0.16	
	ML2	5.74	(0.2)	39	3.48	
	MT	4.44	(1.0)	40	1.19	
	MT2	8.46	(1.0)	40	95.17	
Y(III)	M	–		–	0.06	+9.56
	ML	2.86	(2.0)	38	1.70	
	ML2	5.44	(2.0)	38	24.25	
	ML3	7.20	(2.0)	38	50.03	
	MT	4.03	(0.05)	40	2.84	
	MT2	6.89	(0.1)	40	21.11	
La(III)	M	–		–	0.51	+35.44
	ML	2.62	(0.1)	38	10.60	
	ML2	4.42	(0.1)	38	29.55	
	ML3	5.53	(0.1)	38	14.80	
	MT	3.74	(0.05)	40	12.39	
	M(HT)	2.50	(0.1)	40	0.98	
	MT2	6.13	(0.1)	40	31.17	
Zn(II)	M	–		–	9.64	+41.24
	ML	1.70	(1.0)	38	31.08	
	ML2	2.99	(1.0)	38	28.44	
	ML3	3.40	(1.0)	38	2.50	
	MT	2.20	(0.5)	40	25.02	
	MT2	5.00	(0.0)	40	3.31	

TABLE 6.2 (continued)

Metal ion	Equilibrium	Log β	(IS)	Ref.	Mole percent	Weighted average charge
Cd(II)	M	–		–	29.62	+93.50
	ML	1.24	(1.0)	38	33.11	
	ML2	2.16	(1.0)	38	12.93	
	ML3	2.50	(1.0)	38	0.97	
	MT	1.70	(1.0)	39	21.26	
	M(HT)	0.96	(1.0)	39	2.11	
Ca(II)	M	–		–	42.09	+110.36
	ML	0.92	(1.0)	38	22.52	
	ML2	1.42	(1.0)	38	3.34	
	MT	1.80	(0.2)	40	28.39	
	M(HT)	1.11	(0.2)	40	3.66	
Mg(II)	M	–		–	55.39	+136.89
	ML	0.81	(1.0)	38	23.00	
	ML2	1.47	(1.0)	38	4.93	
	MT	1.36	(0.2)	40	13.57	
	M(HT)	0.92	(0.2)	40	3.11	

Note: L = hydroxyisobutyrate ion, M = metal cation, β = formation constant, IS = ionic strength, β are overall formation constants. T = tartrate for most metals, but T = lactate for the cations of Al, In, and Sc.

multi-charged species. Equation 20 treats retention as a linear function of the charge of the metal species, whereas the true relationship is a power function (eqns. 17, 18). However, eqn. 20 can be used without serious error for those metal–ligand systems which are comprised of low proportions of multi-charged cationic metal species; this includes both free metal cations and metal–ligand species. The analyst ultimately determines the equilibrium distribution of metal species by the choice of complexing agent(s), and their concentrations, for use in the mobile phase.

The effect of multi-charged cationic species on the retention observed for different metals is illustrated in Table 6.3. Here, the ion-exchange capacity of a column (C_{18}) is varied by using different mobile phase concentrations of n-octanesulfonate. Equations (17) and (18) show the relationship between the ion-exchange capacity of a column and the retention of a metal species. The concentrations of complexing agents used throughout these experiments were constant; 58.2 mM α-HIBA and 8.8 mM tartaric acid, pH 4.1.

References pp. 165–166

TABLE 6.3

METAL ION ELUTION ORDER AS A FUNCTION OF n-OCTANESULFONATE CONCENTRATION

n-C$_8$SO$_3^-$ (mM)	Elution order ($\longrightarrow$ increasing retention)										
0.72	Zr	Ga	Sc	Y-Zn		In	Al	La	Cd	Ca	Mg
1.44	Zr	Ga	Sc	Y	Zn-In		Al	La	Cd	Ca	Mg
2.88	Zr	Ga	Sc	Y-In		Al	Zn	Cd	Ca	La	Mg
5.76	Zr	Ga	Sc	In	Al-Y		Zn	Cd	Mg	Ca	La
10.80	Zr	Ga	Sc	In	Al	Y	Zn	Cd	Mg	Ca	La
15.84	Zr	Ga	Sc	In-Al		Y	Zn	Cd	Mg	Ca	La
21.60	Zr	Ga	Sc	Al	In	Y	Zn	Cd	Mg	Ca	La

Note: The mobile phases contained 8.8 mM tartaric acid and 58.2 mM HIBA (pH 4.1). Those metals that coeluted are connected by a hyphen in the table.

The data in Tables 6.1–6.3 illustrate the utility of the weighted average charge calculation for estimating the elution order for samples containing several metal cations. Very good correlation between the experimentally observed and the calculated elution order is possible if the metals to be separated are of the same elemental family, and if formation constants are available for all equilibrium reactions between the cations and the mobile phase ligands. The correlation between calculated and experimentally observed elution order for cations with different chemical and physical properties is also very good. For such diverse ions, however, it is necessary for the analyst to carefully select mobile phase ligands, and use appropriate concentrations in order to minimize the equilibrium distribution of multi-charged cationic species.

6.4. CONCLUSIONS

In this chapter we have tried to overview some of the achievements realized within the past ten years for the HPLC separation of metal species. Complexing agents have played a key role in advancing this field, and the use of complexing eluents for the ion-exchange separation of metal species is currently the HPLC method of choice for most inorganic cation analyses. The development of new detector technology and the design of innovative post-column reaction systems has spurred the development and implementation of many of the novel methods for separating metal species, and this trend will certainly continue with the fruition of new technical advances.

REFERENCES

1   F.A. Cotton and G. Wilkinson, Advanced Inorganic Chemistry, Wiley-Interscience, New York, 4th ed., 1980, p. 66.
2   G.J. Sevenich and J.S. Fritz, Anal. Chem., 55 (1983) 12–16.
3   S. Elchuk and R.M. Cassidy, Anal. Chem., 51 (1979) 1434–1438.
4   I.S. Krull, Reaction Detection in Liquid Chromatography, Marcel Dekker, New York, 1986.
5   M. Blaszkewicz and B. Neidhart, Intern. J. Environ. Anal. Chem., 14 (1983) 11–21.
6   W. Langseth, Talanta, 31 (1984) 975–978.
7   H. Suyani, D. Heitkemper, J. Creed and J. Caruso, Appl. Spec., 43 (1989) 962–967.
8   R.M. Cassidy and S. Elchuk, J. Chrom. Sci., 18 (1980) 217–223.
9   A.D. Kirk and A.K. Hewavitharana, Anal. Chem., 60 (1988) 797–801.
10 A.M. Bond and G.G. Wallace, Anal. Chem., 56 (1984) 2085–2090.
11 R.M. Smith, A.M. Butt and A. Thakur, Analyst, 110 (1985) 35–37.
12 R.M. Smith and L.E. Yankey, Analyst, 107 (1982) 744–748.
13 M.L. Marina, J.C. Diez-Masa and M.V. Dabrio, HRCandCC, 9 (1986) 300–301.
14 J.J. Hagan, S.C. Taylor and M.F. Tweedle, Anal. Chem., 60 (1988) 514–516.
15 H.C. Mehra and W.T. Frankenberger, Jr., Analyst, 114 (1989) 707–710.
16 D.O. Cambell, Proc. 10th Rare Earth Res. Conf., 1973, pp. 1098–1107.
17 A.Kerr, W. Kupferschmidt and M. Attas, Anal. Chem., 60 (1988) 2729–2733.
18 W.G. Lakata, E.P. Lankmayr and K. Muller, Fresenius' Z. Anal. Chem., 319 (1984) 563–568.
19 P. Jones, P.J. Hobbs and L. Ebdon, Analyst, 109 (1984) 703–707.
20 D.R. Jenke and G.K. Pagenkopf, Anal. Chem., 55 (1983) 1168–1169.
21 A.Siriraks, J.E. Girard and P.E. Buell, Anal. Chem., 59 (1987) 2665–2669.
22 R.M. Cassidy and S. Elchuk, J. Liq. Chrom., 4 (1981) 379–398.
23 R.M. Cassidy and S. Elchuk, Anal. Chem., 54 (1982) 1558–1563.
24 P. Jones, P.J. Hobbs and L. Ebdon, Anal. Chimica Acta, 149 (1983) 39–46.
25 G.J. Schmidt and R.P.W. Scott, Analyst, 109 (1984) 997–1002.
26 T. Okada, Anal. Chem. 60 (1988) 2116–2119.
27 E.A. Boyle, B. Handy and A. van Geen, Anal. Chem., 59 (1987) 1499–1503.
28 R.W. Slingsby and J.M. Riviello, LC Magazine, 1 (1983) 354–356.
29 S.S. Heberling, J.M. Riviello, M. Shifen and A.W. Ip, Res. and Dev., Sept. (1987) 74–77.
30 B.D. Karcher and I.S. Krull, J. Chrom. Sci., 25 (1987) 472–478.
31 B.D. Karcher, Ph.D. dissertation, Northeastern University, Boston, MA (1988).
32 E. Lederer and M. Lederer, Chromatography, 2nd. edn., Elsevier, Amsterdam, 1959.
33 J.P. Foley, Chromatography, 5 (1987) 43–52.
34 J.P. Foley and W.E. May, Anal. Chem., 59 (1987) 102–109.
35 J.P. Foley and W.E. May, Anal. Chem., 59 (1987) 110–115.
36 H.S. Rossotti, Talanta, 21 (1974) 809–829.
37 D.G. Peters, J.M. Hayes, and G.M. Hieftje, A Brief Introduction to Modern Chemical Analysis, W.B. Saunders Co., Philadelphia, PA, p. 38.
38 R.M. Smith and A.E. Martell, Critical Stability Constants, Vol. 3, Plenum Press, New York, 1975, pp. 36–38.
39 D.D. Perrin, Stability Constants for Metal–Ion Complexes, Part B, Pergamon Press, New York, 1979.

40   R.M. Smith and A.E. Martell, Critical Stability Constants, Vol. 3, Plenum Press, New York, 1975, pp. 127–129.

I.S. Krull (Ed.), *Trace Metal Analysis and Speciation*
Journal of Chromatography Library Series, Vol. 47
© 1991 Elsevier Science Publishers B.V., Amsterdam

167

Chapter 7

Instrumentation and Procedures for Long-Term Automated Monitoring of Metal Ions in Industrial Effluents by Liquid Chromatography with Electrochemical Detection

A.M. BOND

Department of Chemical and Analytical Sciences, Deakin University, Geelong, Victoria 3217, Australia

7.1. INTRODUCTION

The on-line monitoring of the concentrations of toxic metals such as lead, cadmium and mercury in industrial effluents, river water, drinking water, lake water and marine water has become an important aspect of environmental control. Ideally, in any on-line system for monitoring of industrial effluents which flow into the environment, the sampling, chemical treatment to remove interferences (if required), data manipulation, data reporting and all other necessary operations are completely automated. In the present era of chemical instrumentation, normally this means that all operations from the start of the experiment (sampling) to preparation of the final report are under microprocessor control.

Historically, electrochemical techniques were among the first to be used for the determination of metals in aqueous solutions. In the early applications, a sample was taken and placed into a beaker. Subsequently, reagents were added to the beaker to remove interferences, and elemental determinations were undertaken by some form of potentiometry, amperometry, coulometry, voltammetry (polarography) etc. using a metal-sensing electrode or its equivalent and a reference electrode. Virtually, all the manipulations involved in the procedures relevant to electrochemical detection, including the calculation of concentration and cleaning and maintenance of electrodes, continued to be performed manually in commercially available instruments until the era of digital electronics. Thus, on-line monitoring of metals in industrial effluents, utilising strategies based on electrochemical detection, have remained essentially laboratory based curiosities prior to the advent of inexpensive computers, although it should not be forgotten that automated electroanalytical measurements of pH in harsh industrial conditions, and potassium monitoring in clinical and medicinal chemistry with ion-selective electrodes, have been successful even in the era preceding the microprocessor [1, 2].

As might have been predicted, the wide acceptance of the micro-processor in chemical instrumentation in the 1980s, coupled with the availability of many years of literature devoted to the subject of electro-analytical determination of metals, has led to a vigorous and renewed research interest in the area of automated on-line monitoring of metal concentrations in industrial effluents and in process control. Electro-chemical techniques utilise functions of potential, current, time—or a derivative of one of these parameters—and relate the information gained from such measurements to the metal ion concentration. The techniques are therefore natural companions to the computer, which readily recognises potential (voltage) and time. Measurement of current is as trivial as the measurement of potential with digital electronics. The current signal can be translated to a voltage signal and monitored by measurement of the voltage drop occurring when the current flows across a standard resistor (a current-to-voltage converter).

The above paragraphs contain information which implies that the excellent and long pedigree of electroanalytical chemistry should have ensured its widespread use in automated, on-line monitoring of metal concentration in practical applications once computers and the related power of digital electronics became available. However, most studies in which electrochemically based instruments have been developed, have been confined to the research laboratory. Indeed, the inherently more complicated, expensive and less computer compatible spectro-scopic methods have made a far greater impact in the area of on-line monitoring of chemicals in industrially and environmentally important problems.

The notorious problem with electroanalytical methods concerns the reliability and maintenance of electrodes over lengthy periods of op-eration. Almost every student of analytical chemistry has a legendary story to tell of the failure of his or her electrode at a critical stage of an undergraduate experiment. Electrodes are perceived to be inherently fragile and prone to failure even in an ideal laboratory situation. The undergraduate student, when moving to employment in industry and armed with this experience, is unlikely to contemplate suggesting the introduction of operator unattended, low maintenance, long-term elec-trochemical monitoring of metals in a harsh industrial environment. Additionally, while the generally high sensitivity of many electroanalytical methods for metal determinations has been recognized for a long time, it has generally been accepted as a low resolution technique. Single element capability has often been the limit of achievement with four to six element capacity generally being regarded as an upper limit in exceptional circumstances.

As if unreliability and low resolution were not enough disadvantages relative to spectroscopic methods, then the commonly observed problems of interference resulting from even subtle changes in matrix might dissuade even the electrochemical enthusiast from contemplating long-term monitoring of industrial effluents, where considerable changes in matrix may occur even on an hourly basis. Calibration, using standard solutions, therefore can be presumed to be unreliable, unless an intelligent approach is employed in the data evaluation procedures.

Over the past 10 years, a Deakin University research group working in collaboration with research teams from Australian industry and government laboratories have explored techniques and methodologies for undertaking fully automated long-term monitoring of metal concentrations in industrially and environmentally important situations. Basically, we have been endeavouring to extract and utilise efficiently the low cost, extremely computer compatible and highly sensitive advantages of electrochemical techniques, from the plethora of problems that surround the utilisation of the methodology in a practical situation. Many different electroanalytical methodologies have been pursued in the course of these studies. The only concept reviewed in this paper concerns the development of an instrument known affectionately as the METSCAN. The final version of this instrument, developed after ten years of research, utilises a sequence of automated steps, involves sample preparation from a flowing industrial effluent or river, automated formation of electroactive metal complexes, automated injection of the complexes onto a chromatographic column, chromatographic separation of the complexes, and finally electrochemical detection of the complexes. The chromatography provides a clean-up procedure, minimises matrix effects and provides the desired resolution for multi-element capabilities. The METSCAN system utilises electrochemical cleaning of the electrodes and can be used continuously without operator intervention or routine maintenance for a seven day period.

The METSCAN instrument has evolved through several generations of ideas and prototypes and has been predominantly the result of a long-term collaboration between Deakin University, Geelong, Victoria, Australia and the Ordnance Factory, Maribyrnong, Victoria, Australia. Important contributions from my colleagues (in alphabetical order)

M.J.P. McBurney, E.T. Crosher, W.N.C. Garrard, I.D. Heritage, L.S. McLachlan, T.P. Majewski and G.G. Wallace, in combination with the technical staff of our respective institutions and external consultants, have been required to achieve successfully the long-term objective set 10 years ago, which was to design and build computer based automated liquid chromatographic instrumentation with electrochemical detection for long-term monitoring of metals in industrial effluents. A series of papers, many the results of these endeavours, is available in references 3 to 12. This account of metal monitoring is a review in more or less chronological order of the various approaches used in the prototype instrument leading to the final development of the METSCAN. Experience has taught us both the drawbacks and limitations of many aspects of automation of electrochemical techniques, and it is my hope that a didactic account of both our successes and failures may be of interest to those who wish to pursue this challenging field of research.

7.2. CHROMATOGRAPHY AND ELECTROCHEMISTRY OF METAL COMPLEXES

In any method for metal determination using liquid chromatography with electrochemical (voltammetric) detection, a method for automated preparation of suitable metal complexes must be included in the analytical scheme. The proposed complexes must, of course, be capable of chromatographic separation and oxidation or reduction at the electrochemical detector. Numerous liquid chromatographic studies of metal complexes have been reported over the past few years. Detailed overviews of the field of separation and determination of metals as chelated complexes may be gained by perusal of references 13 to 27. The majority of the studies reported to date have used reverse phase high performance liquid chromatography at bonded stationary phases after pre-column derivatization, although other forms of chromatography, such as size exclusion chromatography, ion chromatography and normal phase chromatography, have also been employed. Frequently used derivatizing agents include dithiocarbamates and related compounds [3–12, 28–39], 8-quinolinol [40–43], β-diketones [44] and 4-(2-pyridylazo)resorcinol [45–52].

When the complexes can be reduced or oxidized so that electrochemical (voltammetric) detection can be employed, the most common method involves amperometric detection at a constant DC potential. With this method, the current, which is proportional to the metal concentration,

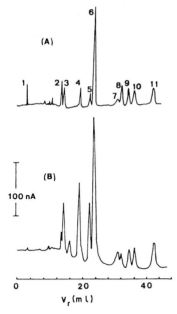

Fig. 7.1. Multi-element determination by reverse phase chromatographic separation of metal dithiocarbamate complexes and (A) spectrophotometric detection at 254 nm and (B) amperometric detection at a glassy carbon electrode with an applied DC potential of +1.20 V vs Ag/AgCl in a thin layer cell. Reproduced by courtesy: *Anal. Chim. Acta* 164 (1984) 223.

is measured at a potential which is sufficiently negative to reduce the metal complex of interest, or sufficiently positive to oxidize it. Figure 7.1 illustrates the readout obtained when metal complexes are formed by pre-column derivatization, separated by reverse phase chromatography and monitored spectrophotometrically or electrochemically by amperometric detection [9]. Traditionally, spectrophotometric detection has been used more widely than electrochemical detection, and has often been used as a reference method for developing techniques described in this review. Indeed, much of what is described with respect to automation applies equally well to both methods of detection.

The electrochemical cells employed for the detection have usually been one of two types: wall jet (fluid impinges directly onto the electrode) or thin layer (fluid flows directly over the electrode surface). Reviews of the technique of liquid chromatography with electrochemical detection [53–56] contain detailed descriptions of the cells used for amperometric and other kinds of electrochemical detection.

The majority of studies utilising chromatographic separation of metals with electrochemical detection reported in the literature, refer to laboratory based determination of metals, where discrete batch methods of sampling from individually prepared samples are employed. In contrast, the concept to be reviewed in detail in this paper concerns continual automated monitoring of industrial effluents (or any stream of solution), over lengthy periods of time without any operator intervention. This review article therefore describes the development of automated on-line monitoring procedures based upon the formation, chromatographic separation and amperometric detection, of metal ions as their dithiocarbamate complexes.

I, dtc⁻   II, dedtc⁻

III, pydtc⁻   IV, hedtc⁻

Structures I–IV (see above) provide the general formula (one resonance form used) for the dithiocarbamate ligand (I), dtc, and the particular examples of diethyldithiocarbamate (II), dedtc⁻, pyrrolidinedithiocarbamate (or pyrrolidinecarbodithioate) (III), pydtc⁻, and bis(2-hydroxyethyl)dithiocarbmate (IV), hedtc⁻, which will be used in this review. The dithiocarbamate ligand complexes with almost every environmentally important metal, [57], is readily and inexpensively synthesised, and the metal complexes almost invariably can be either oxidised or reduced [58]. Importantly, the dithiocarbamate ligand has had a tradition of being an analytically useful reagent for more than a century [59], so a great knowledge base is available from which to draw the required chemical information. Other ligands could be utilised as alternatives to dithiocarbamates, but few would provide the extreme versatility, and at the same time have such a broad and extensive knowledge data base to call upon with respect to methodology development.

7.3. THE OFF-LINE DETERMINATION OF COPPER AS A DITHIOCARBAMATE COMPLEX BY REVERSED-PHASE LIQUID CHROMATOGRAPHY WITH ELECTROCHEMICAL DETECTION

7.3.1. General principles

Before considering the special requirements for automated on-line determination of metals from industrial effluents, it is worthwhile examining the features of standard laboratory procedures associated with the off-line determination of copper as a dithiocarbamate complex by liquid chromatography with electrochemical detection. The off-line determination of copper as its diethyldithiocarbamate complex in aqueous samples, zinc plant electrolyte, and urine have been described [3, 7, 10] using reverse phase liquid chromatography with amperometric detection. A standard instrumental configuration for the conventional laboratory off-line method as used in these studies is depicted in Fig. 7.2.

Literature reports demonstrate that many metal dithiocarbamate complexes are formed by the reactions given in equations (1) and (2):

$$M^{2+} + 2\,dtc^- \rightarrow M(dtc)_2 \tag{1}$$

$$M^{3+} + 3\,dtc^- \rightarrow M(dtc)_3 \tag{2}$$

where M^{2+} is a divalent metal cation such as Cu^{2+}, Ni^{2+} and M^{3+} is a trivalent metal cation such as Mn^{3+}, Co^{3+}, Cr^{3+}. The metal complexes, $M(dtc)_2$ and $M(dtc)_3$ usually are only sparingly soluble in water, but very soluble in organic solvents. This characteristic has made metal dithiocarbamate complexes extremely valuable in developing sensitive analytical methods based on solvent extraction from aqueous media coupled with detection of the metal by atomic absorption spectrometry directly in the organic solvent [60, 61]. Available data in the literature also indicate that metal dithiocarbamate complexes can be obtained in a range of oxidation states in suitable organic solvents, e.g. Mn(IV), Mn(III), Mn(II) Fe(IV), Fe(III), Fe(II) Cu(III), Cu(II), Cu(I) [58]. In the case of the $Cu(dtc)_2$ complex, it would be predicted therefore that both reversible oxidation

$$Cu(II)(dtc)_2 \rightleftharpoons [Cu(III)(dtc)_2]^+ + e^- \tag{3}$$

and reduction

$$Cu(II)(dtc)_2 + e^- \rightleftharpoons [Cu(I)(dtc)_2]^- \tag{4}$$

174

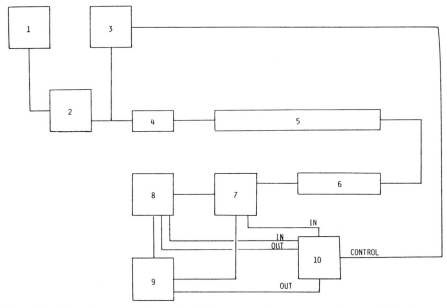

Fig. 7.2. Flow diagram of standard laboratory instrumentation used for chromatography with electrochemical or spectrophotometric detection of metal dithiocarbamate complexes. *1* = chromatographic solvent, *2* = solvent delivery system, *3* = injection system, *4* = guard column, *5* = separator column, *6* = suppressor column, *7* = spectrophotometric detector, *8* = electrochemical detector, *9* = readout device, *10* = microprocessor. Reproduced by courtesy: *J. Liquid Chromatog.* 6 (1983) 1799.

processes would be available for electrochemical detection after extraction into an organic solvent.

Several workers have extracted dithiocarbamate complexes into organic solvents and detected the element electrochemically [62, 63] after procedures involving destruction of the complexes and a back extraction into an aqueous phase. However, direct determination of the Cu(dtc)$_2$ complex in the organic phase is likely to be a simpler method. If reverse-phase chromatography is used to separate the dithiocarbamate complexes, then mixed aqueous–organic solvents such as methanol–water or acetonitrile–water will be the mobile phase and amperometric detection of the dithiocarbamate complexes can occur directly in the chromatographic aqueous non-aqueous solvent mixture. The chromatographic solvent usually includes a buffer to control the pH. If the buffer itself does not provide adequate conductivity, then an indifferent electrolyte such as 0.01 M NaNO$_3$ can be added to the chromatographic solvent for the purposes of electrochemical detection.

7.3.2. Electrochemistry of copper dithiocarbamate complexes in a conventional electrochemical cell

Figure 7.3 shows a cyclic voltammogram obtained at a platinum electrode in acetone (0.1 M Et_4NClO_4) for reduction and oxidation of a $Cu(dtc)_2$ complex in a conventional electrochemical cell. With a stationary solution, asymmetrical peak shaped curves are observed, whereas in a flow-through cell, a sigmoidal shaped steady state voltammogram is more likely to be observed. Two chemically reversible processes, one oxidation and one reduction, corresponding to equations (3) and (4) are observed. In acetonitrile (0.1 M Et_4NClO_4), which is a commonly used organic component of the mobile phase used in reverse phase chromatography, cyclic voltammograms obtained in stationary solutions are also chemically reversible, although at some electrode surfaces, the electron transfer step may be only quasi-reversible [67]. Thus at platinum

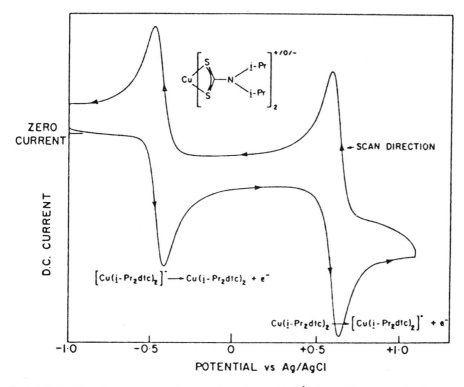

Fig. 7.3. Cyclic voltammogram (scan rate = 200 mV s^{-1}) for oxidation and reduction of $Cu(dtc)_2$ in acetone (0.1 M Et_4NClO_4) at a platinum disc electrode in a stationary cell. Reproduced by courtesy: *Inorg. Chem.* 15 (1976) 2115.

References pp. 210–211

electrodes and employing a scan rate of 200 mV s^{-1}, peak potentials at 20°C in cyclic voltammetry for the Cu(dedtc)$_2$ complex in acetonitrile (0.1 M Et$_4$NClO$_4$), were -0.53 and -0.46 V vs Ag/AgCl for the reduction step and $+0.47$ and $+0.40$ V vs Ag/AgCl for the oxidation process. Amperometric detection of copper as the Cu(dedtc)$_2$ complex in a flow-through cell therefore would be expected to be achieved by applying a constant potential of say $+0.6$ V vs Ag/AgCl or say -0.6 V vs Ag/AgCl if the solvent were acetonitrile and assuming results from static solutions can be transferred to flowing solutions. Results for Cu(pydtc)$_2$and other copper dithiocarbamate complexes at platinum electrodes are essentially the same as with Cu(dedtc)$_2$, except that peak positions are shifted slightly (substituent effect). The cyclic voltammetric responses at gold and glassy carbon electrodes in a conventional cell are similar to those obtained at a platinum electrode. However, at a glassy carbon electrode, the peak-to-peak separations are greater than at Pt or Au, indicating a slower rate of electron transfer at this electrode material. This is generally not important when using amperometric detection in a flowing solution if the constant DC potential is set at a value corresponding to the limiting current region.

Examination of cyclic voltammograms in acetonitrile–water mixtures used in chromatography demonstrate that excellent electrochemical responses for Cu(dtc)$_2$ can still be obtained in the presence of 30% water. However, solubility of the Cu(dedtc)$_2$ and Cu(pydtc)$_2$ complexes decrease as the water concentration increases and the use of water content higher than 30% may result in precipitation of these complexes. With 30% water, up to 4×10^{-4} M concentrations of Cu(dedtc)$_2$ can be used, but with Cu(pydtc)$_2$, solubility is limited to approximately 4×10^{-5} M.

As an alternative to using the pre-prepared Cu(dtc)$_2$ complex, $1:2$ mixtures of copper nitrate and ligand produce electrochemical curves essentially the same as that obtained with the prepared complex. In the presence of a considerable excess of the ligand, Cu(dtc)$_2$ formation is extemely rapid. However, under these conditions, waves due to the oxidation of ligand [58,65,66] mask the Cu(dtc)$_2$ oxidation process. Thus, with a large excess of ligand, only the Cu(dtc)$_2$ reduction wave could be used analytically in a conventional cell.

When reverse-phase liquid chromatography is used with electrochemical detection in a flowing solution, a buffer is usually present in the mobile phase. The buffer itself should be electroinactive and not interfere with the determination of copper. Instead of using 30% water (0.2 M NaNO$_3$), a suitable aqueous component would be say 30% water (0.005 M NaNO$_3$, 0.02 M acetate buffer). The low concentration of buffer ensures the minimum of interference from reduction of hydrogen ion and the pH of the buffer should be compatible with the pH restrictions applicable to many reverse-phase columns. As ascertained by monitoring of differential pulse voltammograms [67, 68] as a function of time, it can be shown that the Cu(dedtc)$_2$ complex is unstable below pH 6, whereas the Cu(pydtc)$_2$ complex is completely stable in the pH range 5–7. In view of the above information, a pH buffer of 6.0 is suitable for chromatographic separation and electrochemical detection of either copper dithiocarbamate complex.

7.3.3. Reverse phase liquid chromatography with electrochemical detection

On the basis of findings from the preliminary investigation of the above kind in a conventional electrochemical cell, a suitable mobile phase for the reverse phase liquid chromatographic separation of Cu(dtc)$_2$ complexes with electrochemical detection would be 70% acetonitrile–30% water (0.02 M acetate buffer) with NaNO$_3$ as supporting electrolyte. Electrodes investigated in the published paper [3] were the same as in the stationary cell and both oxidation and reduction processes for Cu(dtc)$_2$ were compared. A Metrohm EA 1096 detector cell (wall jet electrode) was used in this particular cell and a C$_{18}$ reverse phase chromatographic column was employed. Retention volumes of 14.4 and 10.4 ml were obtained for Cu(dedtc)$_2$ and Cu(pydtc)$_2$, respectively. This smaller retention volume of Cu(pydtc)$_2$ may be attributed to the more polar nature of the complexes. Other experimental details are available in reference 3. In addition, the possibility of *in situ* formation of the Cu(dtc)$_2$ complex as an alternative to *ex situ* formation of the complex externally to the column also was examined.

7.3.4. *Ex-situ* formation of copper dithiocarbamate complexes

The results described in this section refer to Cu(dtc)$_2$ complexes formed externally to the column in the presence of an excess of dtc⁻.

7.3.4.1. *Detection in the reduction mode*

Figure 7.4 shows the chromatographic response for Cu(dtc)$_2$ using the DC detection method (reduction mode) at a glassy carbon electrode. A well-defined response for both Cu(dtc)$_2$ and oxygen is seen. Even with

178

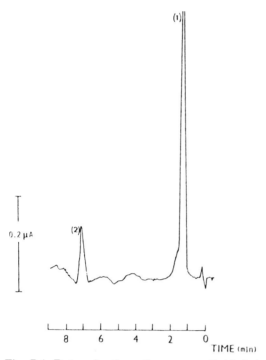

Fig. 7.4. Determination of copper as Cu(dedtc)$_2$ by the reduction process at a glassy carbon electrode and *ex situ* complex formation: (*1*) current response for reduction of oxygen; (*2*) current response for reduction of Cu(dedtc)$_2$. Reverse phase chromatography used with a C$_{18}$ column. DC current measured at a potential of -0.6 V vs Ag/AgCl. Flow rate, 2 ml min^{-1}. Injection volume, 20 μl (200 ng copper). Temperature, $20 \pm 1°$C. Reproduced by courtesy: *Anal. Chem.* 53 (1981) 1209.

extensive endeavours to eliminate oxygen from the injected sample, a residual peak due to reduction of oxygen is normally observed. Partial removal of oxygen from the running solvent via nitrogen gassing is essential to decrease the noise level associated with the glassy carbon electrode response to an acceptable level. A detection limit of 15 ng of copper was obtained by using the reduction process at a glassy carbon electrode direct current (DC) detection and with a 20 μl injection. Use of reverse pulse waveform [69] instead of a constant DC potential further discriminates against detection of the oxygen process, e.g. initial potential -0.45 V, final potential -0.20 V vs Ag/AgCl. In the acetonitrile–aqueous medium, the oxygen reduction process is irreversible and can be discriminated against in favour of the more reversible Cu(dtc)$_2$ reaction when using a pulse method. The pulse technique enables a detection limit of 1 ng of copper to be obtained with a 20 μl injection and provides a significant advantage with respect to both sensitivity and minimisation of the oxygen problem. In the absence of this kind of problem, DC and pulse techniques usually give similar limits of detection in a flow-through cell. At platinum and gold electrodes, the limited potential range available at pH 6 for reduction causes difficulties and there is a greater degree of interference (noise) from the hydrogen ion reduction process. Glassy carbon is therefore the preferred electrode for use with the reduction process, but removal of oxygen is a non-trivial problem.

7.3.4.2. Detection in the oxidation mode

Figure 7.5 shows the chromatographic response obtained at a gold electrode when using the oxidation process and the DC detection technique at a potential of +0.6 V vs Ag/AgCl. The first peak corresponds to oxidation of the excess ligand and occurs near the solvent front. Thus, much of the interference from excess ligand observed in electrochemical methods performed in a conventional cell is eliminated by the chromato-

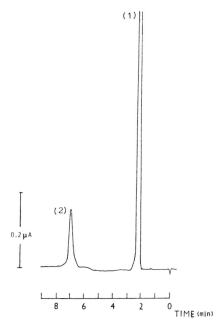

Fig. 7.5. Determination of copper as $Cu(dedtc)_2$ using the oxidation process at a gold electrode and *ex situ* complex formation: (*1*) current response due to oxidation of ligand; (*2*) current response for oxidation of $Cu(dedtc)_2$. DC current measured at +0.6 V vs Ag/AgCl. Other conditions as in Fig. 7.4. Reproduced by courtesy: *Anal. Chem.* **53** (1981) 1209.

References pp. 210–211

TABLE 7.1

DATA OBTAINED BY LIQUID CHROMATOGRAPHY WITH ELECTROCHEMICAL DETECTION OF EITHER Cu(dedtc)$_2$ OR Cu(pydtc)$_2$ [a]

Electrode	Electrode process	Detection limit [b] (ng of Cu)		Concentration range of linear response (ng of Cu)	
		ex situ method	in situ method	ex situ method	in situ method
Glassy carbon	reduction	15	15	15–400	15–400
Glassy carbon	oxidation	10	10	10–700	10–200
Platinum	oxidation	1	1	1–250	1–200
Gold	oxidation	1	1	1–450	1–600

[a] DC current measured at $+0.6$ V vs Ag/AgCl for the oxidation process and -0.6 V vs Ag/AgCl for the reduction process. Flow rate, 2 ml min^{-1}. Injection volume, 20 μl. Temperature, $20 \pm 1°C$. Data taken from reference [3].
[b] For signal-to-noise ratio of 2 : 1.

graphic separation. A similar chromatographic response is obtained at platinum and glassy carbon electrodes. Table 7.1 provides detection limits and the concentration range over which the peak height is a linear function of concentration. The wider linear range of the calibration curve for copper at gold electrodes coupled with the lower detection limit made the oxidation process with this electrode the recommended method for determination of copper using the conditions relevant to the present work. However, for many other elements, requiring the use of more positive potentials for detection and therefore in the general case, glassy carbon is the preferred electrode material. Results are almost the same when using either Cu(dedtc)$_2$, Cu(pydtc)$_2$ or other dithiocarbamate complexes. Little or no advantage in using either the differential or normal pulse techniques is found when using the oxidation process.

Injection volumes of sample and flow rate of the mobile phase are parameters which may be varied under conditions of the preliminary study. Injection volumes greater than 40 μl produce a decrease in the current per unit concentration. Consequently, injection volumes in the range of 20–40 μl are employed in both conventional and automated systems described later in this paper. An increase in flow rate leads to an increase in current because of increased convection. With a conventional C$_{18}$ column, and for flow rates above about 2 ml min^{-1}, the variation of current with flow rate is marginal and the decrease in effective plate count occurring as the flow rate increases means that flow rates in the range of 1–2 ml min^{-1} are optimum for determining copper.

However, the use of higher flow rates means the amount of solvent consumed is greater and in the on-line versions of the intrumentation, the use of the lower flow rates and microbore chromatography will be emphasised.

7.3.5. *In situ* formation of copper dithiocarbamate complexes

As an alternative to generating the complex prior to injection onto the column, *in situ* formation of the complex on the column was examined by including 10^{-3} M ammonium pyrrolidinedithiocarbamate in the mobile phase. Results with this technique are summarized in Table 7.1. Generally, similar detection limits were obtained with *ex situ* or *in situ* complex formation methods (Table 7.1). However, characteristics are not always identical. Figure 7.6 shows the peak height as a function of flow rate with the *in situ* method. The dependence of peak height on flow rate is seen to be different to that for the *ex situ* method (Fig. 7.7). Results from injection volume studies were similar to those obtained with *ex situ* complex formation, and it is again recommended with this complex formation method that injection volumes be kept below 40 μl.

The *in situ* method has some advantages over the *ex situ* method. For example: (i) a real sample can be collected, filtered, and injected onto the column without further sample preparation; (ii) sensitivity is increased in the absolute sense because the dilution factor involved in external complex preparation is eliminated; (iii) there is less chance of decomposition of the complex on the column since excess ligand is

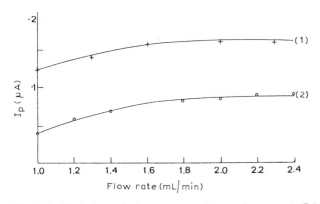

Fig. 7.6. Variation of chromatographic peak current (I_p) with flow rate after forming Cu(pydtc)$_2$ *in situ*: (1) gold or platinum electrodes; (2) glassy carbon electrodes. A total of 60 ng of Cu was injected as copper nitrate. Other conditions as in Fig. 7.5. Reproduced by courtesy: *Anal. Chem.* 53 (1981) 1209.

References pp. 210–211

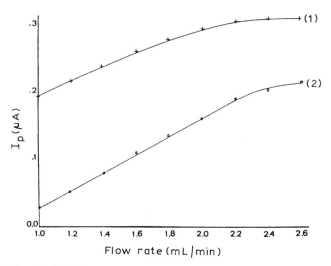

Fig. 7.7. Variation of chromatographic peak current with flow rate using *ex situ* complex formation: (*1*) gold or platinum electrodes; (*2*) glassy carbon electrodes. A total of 200 ng of copper was injected as Cu(dedtc)$_2$. Other conditions as in Fig. 7.5. Reproduced by courtesy: *Anal. Chem.* 53 (1981) 1209.

always present. Disadvantages of the *in situ* method, if the oxidation process is being used, include raising of the background current due to oxidation of the excess ligand present.

7.3.6. Tests for interference

Use of the *in situ* method, with Cu(pydtc)$_2$ oxidation process at a gold electrode, amperometric detection at +0.60 V vs Ag/AgCl, a solvent flow rate of 2 ml min^{-1}, a temperature of 20°C and a C$_{18}$ reverse phase column [3] gives the following results:

(i) *Anions*: The anions NO_3^-, Cl^-, Br^-, I^-, SO_4^{2-}, F^-, Cr_2O_7 and PO_4^{3-} were added separately to a standard 10^{-5} M copper solution. Each anion was added in the form of the sodium or potassium salt in a 10-fold concentration excess. The response for Cu(pydtc)$_2$ complex current was unaltered by the presence of any of the anions examined. The only anion to record a current response at a gold electrode was iodide which came through with the solvent front. Iodide is readily oxidized at gold electrodes.

(ii) *Cations*: K^+, Ca^{2+}, Mg^{2+}, Pb^{2+}, Cd^{2+}, Fe^{2+}, Co^{2+}, Zn^{2+}, Cr^{2+}, Al^{3+}, Mn^{2+}, and Hg^{2+} were added as their chloride or nitrate salts. None of these metals altered the height of the current response for the

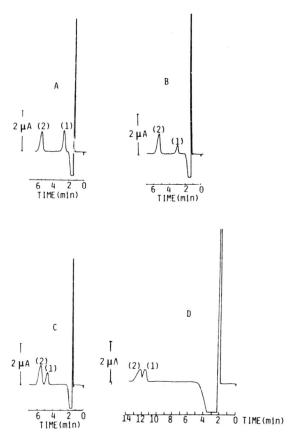

Fig. 7.8. Peaks due to (A) cadmium(II), (B) lead(II), (C) cobalt(III) and (D) iron(III) observed with a gold electrode and forming Cu(pydtc)$_2$ *in situ* for detection of 20 μl of a 5×10^{-4} M copper nitrate solution. Peak 2 corresponds to copper and peak 1 to other elements. Flow rate, 2 ml min^{-1} for A, B and C and 1 ml min^{-1} for D. Other conditions as in Fig. 7.5. Reproduced by courtesy: *Anal. Chem.* 53 (1981) 1209.

copper complex. However, the presence of additional peaks can be noted with some metals as shown in Fig. 7.8. The additional peaks in Fig. 7.8 are due to formation of dithiocarbamate complexes of cadmium(II), lead(II), cobalt(III), and iron(III) with retention volumes of 3.6, 7.5, 9,4 and 11.4 ml, respectively. The iron(III) response overlaps the tail of the copper peak, under the conditions of Fig. 7.8. With a decrease of the flow rate from 2 to 1 ml min^{-1} the copper and iron peaks can be satisfactorily resolved. In the presence of large excesses of iron(III), flow rates need to be carefully adjusted to obtain separation.

The fact that other metals give separate peaks shows that the chromatographic method with electrochemical detection can be used as an attractive method of multielement analysis.

7.3.7. Determination of copper in water samples

All of the above data suggest that a relatively interference-free method for the determination of copper should be possible by liquid chromatographic separation of $Cu(dtc)_2$complexes with electrochemical detection. Determination of real samples over the concentration range 1×10^{-6} to 5×10^{-5} M was undertaken with drinking water in contact with copper pipes and results were compared with those obtained by atomic absorption spectrometry. Excellent agreement, for example, $(2.20 \pm 0.05) \times 10^{-6}$ M and $(14.2 \pm 0.3) \times 10^{-6}$ M vs $(2.20 \pm 0.05) \times 10^{-6}$ M and $(14.3 \pm 0.3) \times 10^{-6}$ M suggests total copper is being determined. Certainly this will be true for atomic absorption spectrometry. Comparisons of data with different techniques on a wide range of industrial effluents are also in excellent agreement [3], so the chromatographic method with electrochemical detection is believed to be relatively specific.

7.4. AUTOMATED ON-LINE DETERMINATION USING OXIDATION PROCESSES FOR METAL DITHIOCARBAMATE COMPLEXES PREPARED IN THE *IN-SITU* MODE

Continuous determination of the concentrations of chemical species is often required in quality control or effluent monitoring programs associated with industrial plants. If it is necessary to collect a sample from the plant, transport it to the laboratory, and finally undertake the determination, the time lag induced between collection and any appropriate action being taken may lead to a costly and inefficient operation. Furthermore, staff may not be available in the analytical laboratory 24 hours a day. The absence of data at these times could lead to delays in detecting a safety or environmental hazard.

The first METSCAN instrument developed for automated monitoring of metals in industrial effluents was based on *in situ* complex formation of rapidly formed metal dithiocarbamate complexes. Complex formation was achieved by including dtc^- in the mobile phase and a schematic diagram of the on-line instrument is presented in Fig. 7.9. Automatic injection and extensive use of microprocessors for experiment control distinguish the on-line system relative to the off-line method described

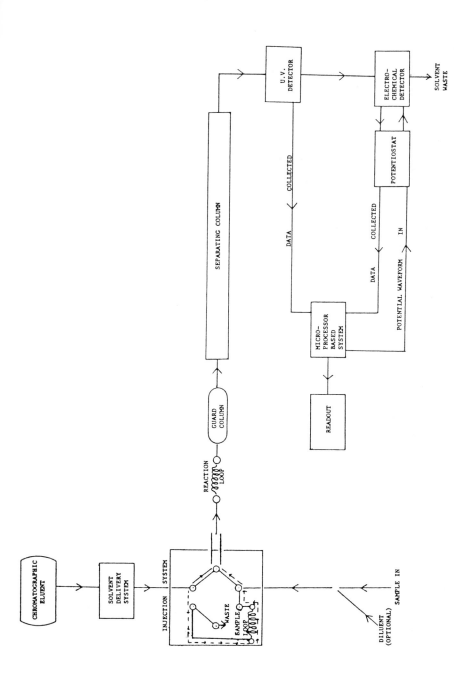

Fig. 7.9. Schematic diagram of the initial version of the automated METSCAN system used for use with *in situ* complex formation. Reproduced by courtesy: *Anal. Chem.* 55 (1983) 718.

References pp. 210–211

in the previous section. Glassy carbon electrodes were employed as the electrode material of choice and oxidation of dithiocarbamate complexes was used rather than reduction. The glassy carbon electrode is more universally applicable than platinum or gold electrodes (see later) and the use of oxidation rather than reduction processes obviates the need to remove oxygen. This latter feature is a distinct advantage in automated monitoring systems.

In the standard laboratory-based form of the instrumentation, a Model 6000A solvent delivery pump was employed in conjunction with a Model U6K universal injector both obtained from Waters Associates. In the first version of an automated system, the solvent delivery pump remained, but the manual injector was replaced by its automatic counterpart, the Model 7126 Automatic Sample Injector from Rheodyne.

The "reaction coil" is 20 cm of coiled stainless steel capillary tubing. The guard column (1.5 cm length, inside diameter) which protects the more expensive C_{18} reversed-phase separating column (length = 30 cm, inside diameter = 3.9 mm) is packed with silica-based C_{18} material. Typical electrochemical detector cells which can be used are (i) Model TL-5 thin layer detector cell obtained from Bioanalytical Systems, (ii) Model EA 1096 wall jet detector cell obtained from Metrohm.

The microprocessor generates all wave forms for applying the potential, collects, stores and calculates data from the results, and is used to control all other steps of the experiment.

7.4.1. On line determination of copper and nickel with *in situ* complex formation

The mobile phase composition 70% : 30% acetonitrile : acetate buffer (0.02 M, pH 6) gives adequate separation of $Cu(dtc)_2$ and $Ni(dtc)_2$. 0.005–0.01 M $NaNO_3$ is added to the solvent to provide adequate conductivity for electrochemical detection. The mobile phase also contains between 10^{-3} and 10^{-4} M dtc^- for the purposes of *in situ* complex formation. Addition of ligand to the solvent does not affect the retention volumes of these compounds.

Both copper and nickel react very rapidly with either $dedtc^-$ or $pydtc^-$ to form the corresponding metal–dithiocarbamate complex. Work in a conventional electrochemical cell as described previously, indicated that a 10-fold excess of ligand was required for fast and complete complex formation. Under these conditions a short length of coiled tubing (20 cm), inserted between the injector and the guard column, allows sufficient time for complete reaction.

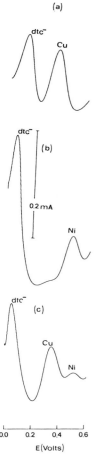

Fig. 7.10. Differential pulse voltammogram in a conventional cell showing the oxidation of $M(dtc)_2$ at a glassy carbon electrode. Delay between pulses, 0.5 s. Pulse amplitude, 100 mV. Scan rate, 5 mV s^{-1}. (a) 5 ppm copper added to 5×10^{-4} M dedtc$^-$ in 70% : 30% acetonitrile : acetate buffer (0.02 M) pH 6, (0.1 M NaNO$_3$), (b) 5 ppm nickel added to the dedtc$^-$ solution, (c) 5 ppm nickel + 5 ppm copper added to the dedtc$^-$ solution. Reproduced by courtesy: *Anal. Chem.* 55 (1983) 718.

Figure 7.10, shows differential pulse voltammograms for the oxidation of $Cu(dtc)_2$ and $Ni(dtc)_2$ in a conventional cell in the presence of excess ligand. These conditions resemble the solution conditions present in the automated system but without the chromatographic separation. While it would be possible to simultaneously determine copper and nickel in a conventional cell under some circumstances, only limited concentration ratios would be available unless chromatographic separation is included.

References pp. 210–211

At around +0.1 V vs Ag/AgCl, oxidation of the free ligand occurs to produce thiuram disulfide.

At more positive potentials, the previously described reversible $Cu(dtc)_2 \rightleftharpoons [Cu(dtc)_2]^+ + e^-$ process and the irreversible $Ni(dtc)_2c$ oxidation process given in equation (5) are observed.

$$3\,Ni(dtc)_2 \rightarrow 2\,[Ni(dtc)_3]^+ + Ni^{2+} + 4\,e^- \tag{5}$$

Recording of voltammograms in the flow-through electrochemical cell, after chromatographic separation, gave the expected results. That is, a $M(dtc)_2$ oxidation response appeared at the retention volume of the complex.

In view of the different oxidation potentials and oxidation mechanisms, the application of different DC potentials, or different types of pulse wave form, will produce varying sensitivity and selectivity for each element. For example, the following potential formats generated from the microprocessor based function generator could be used: (i) constant DC potential [+0.20 V vs Ag/AgCl], (ii) normal pulse waveform [potential from −0.10 to +0.25 V vs Ag/AgCl], (iii) constant DC potential [+0.75 V vs Ag/AgCl] and (iv) differential pulse waveform with an amplitude of 80 mV [+0.40 to +0.48 V vs Ag/AgCl]. Wave form (i) produces a relatively low background current since the oxidation of free ligand occurs to only a limited extent. This free ligand response will be perturbed by metal complex formation giving rise to an indirect method of metal determination. Wave form (ii) has an even lower background current since at the rest potential *no* oxidation of free ligand occurs. Wave form (iii) gives a response resulting from oxidation of the metal complexes and is sensitive to both nickel and copper. Figure 7.11 illustrates this situation. Wave form (iv) is chosen to specifically determine copper. Even in the presence of extremely large excesses of nickel, only a response for copper is observed.

Figure 7.12 shows the readout obtained with an industrial effluent. Figure 7.13 shows typical results from samples obtained from the purification plant of a copper refinery. A large excess of nickel is present at this stage of the plant purification procedures. Use of the solvent 70%:30% acetonitrile:aqueous acetate buffer makes accurate determination of copper difficult with the very high nickel concentrations using spectrophotometric detection (Fig. 7.13a) and some forms of detection (Fig. 7.13b, (i) and (ii)). Use of a different solvent (e.g., 55%:45% acetonitrile:buffer or 70%:30% methanol:buffer) provides adequate separation (Fig. 7.13c). However, an alternative approach to this problem is the use of a differential pulse wave form. Figure 7.13b (iii)

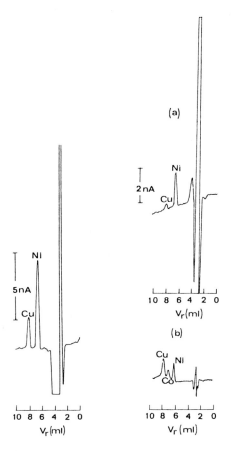

Fig. 7.11 (left). Chromatogram showing separation of nickel and copper as dithiocarbamate complexes formed by the *in situ* method: Chromatographic eluent, 70% : 30% acetonitrile : acetate buffer (0.02 M) pH 6, 0.005 M NaNO$_3$, 10^{-4} M dedtc$^-$. Flow rate, 1 ml min^{-1}. Detection, DC response monitored at +0.75 V vs Ag/AgCl. Injection, 10 μl of sample containing 2 ppm of nickel and 2 ppm of copper. Reproduced by courtesy: *Anal. Chem.* 55 (1983) 718.

Fig. 7.12 (right). Determination of copper and nickel by liquid chromatography with electrochemical injection in an industrial effluent sample. Chromatographic conditions as in Fig. 7.11. Reproduced by courtesy: *Anal. Chem.* 55 (1983) 718.

shows the differential pulse output for a sample with very high nickel concentration which produces a clearly defined copper response void of any interference from the nickel response. This example confirms that if complete resolution is not obtained with the chromatographic stage, then added selectivity can be obtained by "tuning" the detector(s) and/or solvent composition.

References pp. 210–211

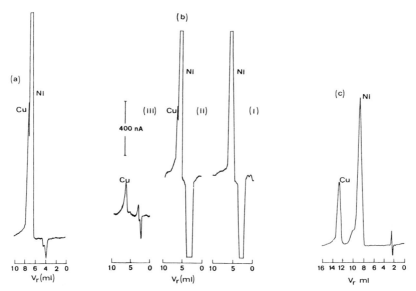

Fig. 7.13. Determination of copper obtained from the purification plant of a copper refinery using liquid chromatography. (a) UV detection ($\lambda = 420$ nm, solvent and other chromatographic conditions as in Fig. 7.11, injection volume = 10 μl), (b) electrochemical detection applying a pulse wave form (initial potential = 400 mV vs Ag/AgCl, final potential = 480 mV vs Ag/AgCl, delay between pulses = 1 s, pulse duration = 0.40 s, (*i*) DC component, (*ii*) pulse component, (*iii*) differential pulse component, injection volume = 10 μl), (c) UV detection ($\lambda = 420$ nm, same conditions as in Fig. 7.11 except that acetonitrile has been replaced by methanol and the injection volume is 3 μl). Reproduced by courtesy: *Anal. Chem.* 55 (1983) 718.

Results for copper and nickel can be obtained at 6 minute intervals in a typical experiment using chromatographic separation and *in situ* prepared dithiocarbamate complexes. Results were also compared with data from off-line experiments using atomic absorption spectrometry and excellent agreement is observed [5].

7.4.2. On-line determination of lead, mercury, cadmium and cobalt with *in situ* complex formation

Lead, mercury, cadmium and cobalt dithiocarbamate complexes are all oxidized at very much more positive potentials than Cu(dtc)$_2$. A potential of +0.95 V vs Ag/AgCl can be used for the determination of cobalt. Careful selection of the applied potential, E_{app}, to avoid the ligand-based thiuram disulfide (background) response enables the determination of lead ($E_{app} = +0.80$ V vs Ag/AgCl) and cadmium ($E_{app} = +0.90$ V vs

Ag/AgCl). However, application of these potentials which are less positive than the oxidation potential, E_{pox}, for these complexes, results in a decrease in sensitivity, relative to that theoretically obtainable if the ligand oxidation process was not present. To achieve adequate sensitivity for these metals, it was also found to be necessary to keep the concentration of dithiocarbamate ligand in the chromatographic solvent below 10^{-4} M. Slightly better signal to noise ratios were obtained when the pydtc$^-$ rather than the dedtc$^-$ ligand was employed for the determination of lead, cadmium or cobalt. However, detection limits for lead and cadmium are relatively high compared to those obtained for copper and nickel, where substantially less positive potentials than that required for oxidation of thiuram disulfide can be applied.

The more positive potential (+1.20 V vs Ag/AgCl) required for oxidation of Hg(dtc)$_2$ precluded its determination using *in situ* formation unless a suppressor column was employed. The function of the suppressor column, which was packed with a commercially available anion exchange resin (Amberlite CG-400), is to remove excess dithiocarbamate ligand after separation but prior to detection. The suppressor column performed this operation very successfully and a sensitive, reproducible (over a period of 25 injections) Hg(dtc)$_2$ response was observed.

The suppressor column, as well as enabling mercury to be determined, was also advantageous with respect to determination of copper, nickel, and in particular cobalt. Unfortunately, use of the suppressor column is detrimental to the determination of Cd(dtc)$_2$ whose response disappears and Pb(dtc)$_2$, whose response is considerably reduced. Presumably, these complexes are significantly retained as anionic complexes on the ion exchange suppressor. These data support conclusions obtained from voltammetric data in a conventional electrochemical cell, that the cadmium and lead complexes exist in anionic as well as neutral forms in the chromatographic solvent. Limits of detection for the simultaneous determination of lead, cadmium, mercury, copper, nickel and cobalt are given in Table 7.2.

Replacement of acetonitrile with methanol in the chromatographic solvent increased the sensitivity of the system for the detection of cadmium. This enhanced response in methanol-containing solvent occurred because Cd(dtc)$_2$ was retained slightly longer. Use of methanol has the added advantage of increasing the resolving power of the chromatographic system with respect to separation of copper and mercury dithiocarbamates. However, unfortunately methanol decreases the overall efficiency of the chromatographic system which, in turn, decreases the capability for multielement determinations. A solvent composi-

TABLE 7.2

DETECTION LIMITS AND REGIONS OF LINEARITY FOR THE DETERMINATION OF LEAD, CADMIUM, MERCURY, COPPER, NICKEL AND COBALT WITH *IN SITU* DITHIOCARBAMATE COMPLEX FORMATION, CHROMATOGRAPHIC SEPARATION AND ELECTROCHEMICAL DETECTION [a]

Element	Detection limit [b,c] (ng)	Linear range [d] (ng)
Lead	20.0	20 –100
Cadmium	10.0	10 – 20
Mercury	0.2	0.2–100
Copper	0.2	0.2–100
Nickel	0.1	0.5–100
Cobalt	0.5	0.5-100

[a] Results were similar using either 10^{-4} M pydtc$^-$ or dedtc$^-$. Injection volume, 10 μl. Further details are available in reference [8].
[b] Signal-to-noise ratio 2 : 1.
[c] For cadmium and lead the detection potential was +0.90 V and +0.80 V vs Ag/AgCl respectively, and no suppressor column was used. For mercury and cobalt the detection potential was +1.20 V vs A/AgCl with the suppressor column being used. For nickel and copper, results were identical using a detection potential of +0.70 V vs Ag/AgCl (no suppressor) or +1.20 V vs Ag/AgCl (with suppressor).
[d] Linearity for reproducibility of better than ±5%.

tion which gives adequate resolution and sacrifices little efficiency is 50 : 20 : 30 acetronitrile : methanol : buffer, when a suppressor column is employed.

7.4.3. Automated multielement analysis over extended periods of time

Long-time monitoring imposes different constraints to short-term measurements. For example, different columns listed below display different performance characteristics with respect to long-term use.

Silica based C_{18} reversed-phase chromatographic columns were used as follows: (i) A C_{18} μBondapak column from Waters Associates, length = 30 cm, internal diameter (i.d.) = 3.9 mm, particle size = 10 μm; (ii) A C_{18} Spherisorb column from Altex, length = 25 cm, i.d. = 4.6 mm, particle size = 5 μm; (iii) A C_{18} Radpak column used in conjunction with a Radial compression module (RCM-100) from Waters Associates, length = 10 cm, i.d. = 8 mm, particle size = 10 μm.

Column iii, the C_{18} Radpak column, gave lower background levels but was found unsuitable for long-term use, because the column back pressure and hence chromatographic efficiency was found to drop throughout the day. Columns i, C_{18} μBondapak, or ii, C_{18} Spherisorb, were found to be suitable for automated analysis over a period of several days, with column (ii) offering superior chromatographic efficiency. With these columns, the automated microprocessor-based system operated continuously with excellent reproducibility over long periods encompasing some 600 injections. Pydtc$^-$ was chosen over dedtc$^-$ for inclusion in the chromatographic solvent for long-term use, because it is more stable.

With electrochemical detection of copper and nickel dithiocarbamates (E_{app} = +0.70 V vs Ag/AgCl), the quality of the responses recorded was found to degrade with time. After 24 hours (200 injections) significant tailing was noted, which indicated that gradual contamination of the electrode surface occurs. This problem was alleviated by use of an electrode cleaning potential pulse (ECLEAN), which was predetermined by the operator and applied by the microprocessor. The potential pulse was applied to the electrode while the injector was in the LOAD mode, for a duration of TFLUSH. An ECLEAN value of -1.20 V vs Ag/AgCl and a TFLUSH value of 30 s were generally used. Over a 72 hour period, the determination of nickel and copper with electrochemical detection was as reproducible as when spectrophotometric detection was used over this period (Fig. 7.14). Detection of lead and cadmium dithiocarbamates was possible with E_{app} = 0.90 V vs Ag/AgCl. However, contamination of the electrode surface, due to thiuram disulfide oxidation, was accelerated dramatically with the use of the more positive potential. In summary, simultaneous determination of all six metals over longer periods of time is confined to determinations using spectrophotometric detection with the *in situ* complex formation method (Fig. 7.14).

Inclusion of the suppressor column while decreasing detection limits for nickel, copper, cobalt and mercury mitigates against long-term use. Generally, the suppressor became saturated within 5 hours (40 injections). In particular, the mercury response indicated signs of degradation after only 20 injections. The lifetime of the suppressor column was found to depend upon the number and composition of samples injected into the system and not just on the concentration of the ligand. It is recommended for specific use, rather than long-term monitoring.

Examination of a range of industrial effluents was considered over an extended time period (72 hours), using electrochemical detection (for Ni and Cu) and spectrophotometric detection (for Ni, Cu, Cd, Pb, Co and Hg). With three sample injections per hour, followed by one standard solution injection, results were reproducible ($\pm3\%$) over a period of

References pp. 210–211

194

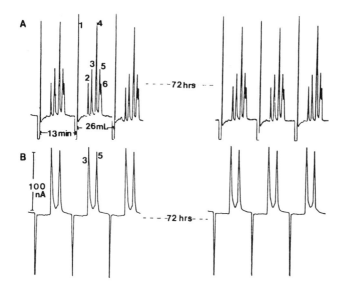

Fig. 7.14. Automated multielement determination using liquid chromatography with *in situ* complex formation over extended time periods using a C_{18} Spherisorb separator column. Chromatographic solvent, 50% acetonitrile : 20% methanol : 30% aqueous buffer and 10^{-4} M pydtc$^-$. Flow rate, 2 ml min^{-1}. Injection volume, 20 μl containing 50 ng of each metal in aqueous solution. (A) Spectrophotometric detection (λ = 254 nm), (B) Electrochemical detection (detection potential, $+0.75$ V vs Ag/AgCl). 1 = Cd(pydtc)$_2$, 2 = Pb(pydtc)$_2$, 3 = Ni(pydtc)$_2$, 4 = Co(pydtc)$_3$, 5 = Cu(pydtc)$_2$, 6 = Hg(pydtc)$_2$. Reproduced by courtesy: *Anal. Chem.* 56 (1984) 2055.

72 hours. Results were the same, within experimental error ($\pm 5\%$) to those obtained by using manual injection or with atomic absorption spectrometry in an off-line mode.

7.5. INSTRUMENTATION FOR 7-DAY CONTINUOUS MONITORING OF METALS USING *EX-SITU* COMPLEX FORMATION

The neutral metal dithiocarbamate complexes may be prepared by two procedures which can be described as *in situ* or *ex situ*. With the *in situ* method, which formed the basis of the first METSCAN instrument, the ligand is included in the chromatographic solvent. At first glance this appears to be the simplest approach to automate and forms the basis of the instrumentation described in the previous section. However, this method has several distinct drawbacks. For example:

(a) The dithiocarbamate ligand, dtc$^-$, is itself oxidized to thiuram disulfide at potentials over the range of approximately 0.2 to 0.8 V vs

Ag/AgCl so that if amperometric detection is used at positive potentials, the electrode is being continuously contaminated by ligand oxidation products. The background current in this potential region is sufficiently raised to accentuate minor fluctuations in flow conditions, thus precluding the use of relatively cheap single piston pumps. At potentials more positive than +0.8 V vs Ag/AgCl a second irreversible ligand oxidation process interferes with the detection of some metal dithiocarbamate complexes oxidised at very positive potentials.

(b) The electrochemical detector response when continuously in the presence of the oxidizable ligand deteriorates after 72 hours. Consequently, a cleaning strategy must be incorporated to restore the performance of the electrode.

(c) Dithiocarbamate ligands such as the diethyl or pyrrolidine derivatives are not completely stable in the mixed aqueous–organic chromatographic solvent and require replacement after approximately 24 to 72 hours.

Ex situ complex formation, while more difficult to automate, eliminates or minimizes many of the above difficulties. *Ex situ* complex formation relies upon external preparation of the metal complex before injection into the chromatographic system. The ligand may therefore be stored in an organic solvent such as acetonitrile and subsequently mixed with the aqueous sample. The dithiocarbamate ligand is considerably more stable in organic solvents than in aqueous media, so that stock solutions can be stored for lengthy periods, particularly when light is excluded.

However, several problems, additional to the greater complexity of automation and which are unique to the *ex situ* method, need careful attention. They are:

(A) Mixing of acetonitrile (or other organic solvent) and water may result in the formation of air bubbles, which if transferred to the chromatographic system, obviously may degrade the system performance.

(B) The mechanics of mixing two solutions thoroughly, allowing time for reaction, and finally, injection into the chromatographic system need to be carefully evaluated and controlled.

(C) Excess dtc⁻ required to ensure complete complex formation on solutions of unknown total metal ion concentration can interfere in a manner similar to *in situ* formation.

A new kind of sample preparation and sample introduction has been developed to overcome problems with *ex situ* formation and is included in the second generation METSCAN instrument now being used in these laboratories. Problems (A) and (B) are overcome by using a

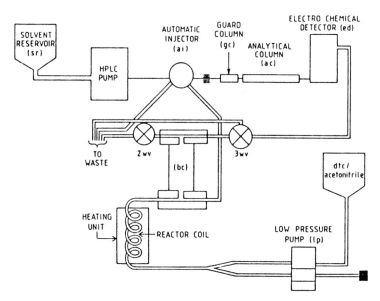

Fig. 7.15. Schematic diagram of the second generation METSCAN system used for trace metal determinations with *ex situ* formation of metal dithiocarbamate complexes. Reproduced by courtesy: *Anal. Chem.* 60 (1988) 1357.

unique reactor vessel and bubble chamber, and problem (C) by the use of an anion exchange guard column to remove unreacted dtc⁻.

The process by which the sample preparation and introduction system works in conjunction with the chromatographic system in the new METSCAN instrument is described via the aid of symbols presented in Fig. 7.15 in the following manner:

The low pressure pumping system (lp) brings together, via a "Y" piece, the filtered test solution and dtc⁻ (in acetonitrile) reagent. This mixture moves through a string bead reactor coil (rc) packed with 3 mm diameter glass balls to achieve complete mixing. The mixing chamber is made by milling a cylindrical chamber (3 cm (l) × 1.5 cm (d)) in an aluminium block, (3 cm × 2.5 cm (b) × 4.0 cm (d)). A heating element and thermostat were mounted on the external surfaces of the block. The solution then passes through the bubble capture/reactor chamber (bc), flushing out this vessel and venting to waste through a valve (2wv). After completing the flush, lp is switched off and the sample in the bubble chamber is allowed time (approximately 5 minutes) to complete the reaction and to allow fortuitous air bubbles to collect at the top of the chamber. The heating unit is employed if the kinetics of complex formation are extremely slow.

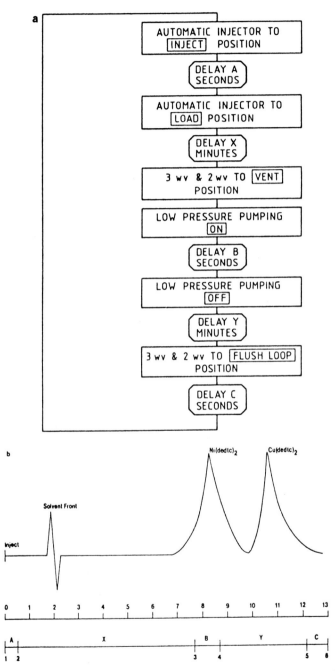

Fig. 7.16. Flow diagram (a) and schematic representation (b) of operational procedures required to produce automatic sample injection and *ex situ* complex formation. Reproduced by courtesy: *Anal. Chem.* 60 (1988) 1357.

At the end of the reaction time, valves 3wv and 2wv are switched so that the chromatographic eluant is directed from the waste to the bubble chamber, thereby pushing the reaction solution out of this chamber into the auto-inject (ai) loop. At the appropriate time during the bubble chamber flush to the automatic injection loop, the injector is switched to the INJECT mode and the sample is introduced into the chromatographic stream. Excess ligand is removed from this stream by use of a high capacity anion exchange guard column (gc), and the neutral complexes are then separated on the analytical column (ac) and detected at the electrochemical detector (ed).

The reaction chamber mentioned above is an essential development which provides a suitable vessel to allow metal complex formation to take place outside the chromatographic part of the system. This chamber consists of a vertical mounted cylinder (32 mm (l) × 7 mm (d)) constructed from polyethylene. An inlet from the mixing chamber and an outlet to the injector are located at the bottom of the cylinder. Connections to the 3-way and 2-way valves were made via a T-piece in the top of the cylinder. The design ensures that bubbles formed as an unavoidable consequence of mixing acetonitrile dtc$^-$ and an aqueous sample, do not pass into the injection part of the chromatographic system at the time of injection. This is accomplished by the entrainment of any bubbles in the upper section of the chamber. Upon flow reversal through the chamber there exists approximately 0.5 ml of "bubble-free" reaction mixture which is flushed out of the chamber by the pumping. Potentiostat control, data acquisition and data evaluation tasks are also undertaken with the aid of the JED computer system. Recommended automated electrode cleaning and calibration procedures are those described previously.

The entire sample preparation and chromatographic system is enclosed in a metal case. Figure 7.17 shows a photograph of the MET-SCAN system in its present form. The potentiostat and electrochemical cell are a novel integrated system built into a single package and contained in a metal case which acts as a Faraday cage (Fig. 7.18). The connecting leads between the potentiostat and cell are less than 1 cm in length, as both are mounted on the same board. The integrated design leads to low noise levels of less than 100 pA (peak-to-peak). A Bioanalytical Systems (BAS) Thin Layer Cell with a BAS glassy carbon working electrode, Ag/AgCl (3 M NaCl) reference electrode and glassy carbon counter electrode was used as the electrochemical cell. The potentiostat was constructed in these laboratories, as was the automatic injector which has been newly developed to achieve long-term reliability. Initial

Fig. 7.17. The METSCAN instrument in its present form (external view). The timing and control of individual components and events to produce automatic sample injection is co-ordinated by a JED computer system based upon a National Semiconductor NSC 800-4 CPU running "tiny BASIC". A flow diagram of the operational procedures is given in Fig. 7.16. Relay switches, driven by the output port, control the ± 12 V solenoid activated valves (2 wv and 3 wv) plus the three 6 volt DC motors used for automatic injection and for the low pressure pumps.

experiments with a standard Rheodyne 7125 injector connected to a 6 volt DC motor via a direct drive did not achieve the required reliability. However, insertion of a simple gear mechanism (along with microswitch position detectors and appropriate electronics) between the motor and the Rheodyne 7125 injector resulted in a reliable automatic injector, which could be operated manually or under computer control.

The solvent reservoir for the METSCAN system holds approximately 2.5 litres and is designed for an unattended seven day continuous cycle period with a flow rate of 0.3 to 0.1 ml min^{-1}. This reservoir is constructed from an inert material. Provision is also made for the collection and storage of waste solvent.

The high pressure chromatographic pump has to produce low flow rates of 0.1 to 0.3 ml min^{-1} with a minimum of flow fluctuation. A Modified ETP-KORTEC (Sydney, Australia) K25M pump with a combined

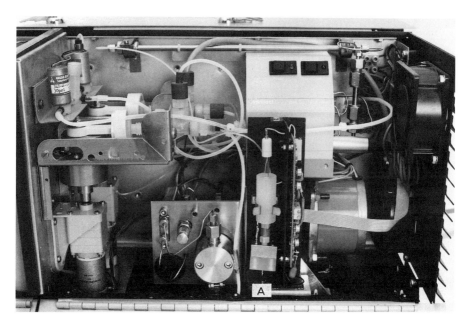

Fig. 7.18. The METSCAN instrument in its present form (internal view) where A is the potentiostat.

pulse dampener/pressure transducer package is used for this purpose. The interfacing of this pump to the JED computer system is simplified by the "on board" logic flow control.

The anion exchange guard column is a disposable Brownlee HPLC cartridge (3 cm × 2 mm) containing high capacity 10 μm anion exchange material. This column removes the excess dithiocarbamate ligand and protects the more expensive analytical column. A high efficiency reverse phase Alltech 250 mm × 2.1 mm, 5 μm Adsorbosphere microbore analytical column is used at flow rates down to 0.1 ml min^{-1} to give retention times of less than 15 minutes for the metal dithiocarbamate complexes of interest.

Two LINK bellows metering pumps combined with 6 volt DC motors have been used to pump the sample and reagent solutions. Teflon tubing or other inert material which does not stretch with time in the presence of the organic solvent is required for transfer of solutions.

In order to achieve a major objective of the METSCAN instrument, which is low solvent flow rate over a seven day unattended operating period, it is essential that microbore chromatography columns should be used instead of the more conventional columns.

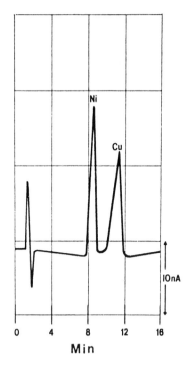

Fig. 7.19. METSCAN determination of 450 ppb Cu(dedtc)$_2$ and 250 ppb Ni(dedtc)$_2$. Solvent, 75% acetonitrile: 25% 0.02 M acetate buffer (pH 6.0), 0.01 M NaNO$_3$. Flow rate, 0.2 ml min^{-1}. Injection volume, 5 μl. Column, Alltech 5 μm Adsorbosphere (25 cm × 2 mm). Detection potential, +0.8 V vs Ag/AgCl. Temperature, 20°C. Reproduced by courtesy: *Anal. Chem.* 60 (1988) 1357.

Adequate resolution at low flow rates is achieved with an Alltech 25 cm × 2 mm, 5 μm C$_{18}$ Adsorbosphere Chromatographic Columns (plate count of approximately 10,000). Figure 7.19 shows the excellent resolution of Cu(dedtc)$_2$ and Ni(dedtc)$_2$ obtained at a flow rate of 0.2 ml min^{-1}. A limit of detection for both metals of less than 10 ppb (signal-to-noise ratio of 2:1) can be achieved with this detection system (at a flow rate of 0.3 to 0.1 ml min^{-1}). Linearity of the peak height versus concentration response was observed for the concentration range 10 to 1000 ppb. The instrumentation described in this section has now achieved the objectives of complete automation for metal determinations, excellent sensitivity, relatively low consumption of organic solvent, and a design compatible with unattended continuous operation for extended periods. Long-term testing of the METSCAN system has been undertaken on the determination of copper and nickel in industrial effluents. With both copper and nickel, a reproducibility of ±5% (average deviation for 100 samples) was obtained routinely at the 100 ppb level over seven day operating periods, when calibration with a standard solution occurs on a daily basis. Determinations were made every fifteen minutes in these trials. This 15 minute period represents the maximum sampling frequency available with the Alltech column and with a flow rate of 0.1 ml per minute. These operating conditions produce no detectable sample carryover. Calibration in the form of injection of a standard reference solution was required every 24 hours to correct for a small decrease in electrode sensitivity and/or reagent decomposition of 3% per day. At the end of the seven day period, complete replacement of solvents, reagents and electrode maintenance is undertaken to restore the detection system to its original operating condition.

7.6. LONG-TERM MONITORING OF METAL CONCENTRATIONS VIA COMPLEXES FORMED BY EXCHANGE PROCESSES WITH A WATER SOLUBLE ZINC DITHIOCARBAMATE COMPLEX

Complete automation of liquid chromatography with an electrochemical detection method has been achieved with the instrumentation described in the previous section, and continual monitoring of metal ions in industrial effluents has been undertaken for seven day periods without operator intervention. In the previous section, diethyldithiocarbamate or pyrrolidinedithiocarbamate complexes of metal ions were prepared automatically in a reactor coil, using a low pressure pumping system which mixed the industrial effluent and an acetonitrile solution containing the dithiocarbamate ligand. The acetonitrile–water solvent mixture formed in the complex preparation stage was then used as the solvent medium for the chromatographic separation of the metal dithiocarbamate complexes via the use of a high pressure pump. Electrochemical detection at a glassy carbon electrode was achieved by oxidation of the metal complex or ligand attached to the metal. After extensive experience with this automated system it was recognized that deficiencies in the completely automated procedure arise from the slow decomposition of the dithiocarbamate ligand over the seven day operation period, the necessary presence of excess electroactive dithiocarbamate ligand which has to be removed prior to electrochemical detection by an anion exchange column, and the use of the relatively expensive and toxic acetonitrile solvent required for the preparation of the water insoluble dithiocarbamate complexes.

In the final variation of the METSCAN instrument, the use of the water soluble, complexing agent, zinc bis(2-hydroxyethyl)dithiocarbamate Zn(hedtc)$_2$, is used as an alternative to salts of diethyldithiocarbamate or pyrrolidinedithiocarbamate. Zn(hedtc)$_2$ has a low stability constant relative to other metals of interest and participates in exchange reactions with many divalent (M^{2+}) and trivalent (M^{3+}) metal ions (equations 7 and 8).

$$Zn(hedtc)_2 + M^{2+} \rightarrow M(hedtc)_2 + Zn^{2+} \tag{7}$$

$$3\,Zn(hedtc)_2 + 2\,M^{3+} \rightarrow 2\,M(hedtc)_3 + 3\,Zn^{2+} \tag{8}$$

The Zn(hedtc)$_2$ complex is water soluble and unlike the previously used dithiocarbamate salts, can be stored in aqueous media for many weeks without decomposition. Additionally, the Zn(hedtc)$_2$ complex is more difficult to oxidize than the free ligand and does not need to be removed from the solution by ion chromatography prior to electrochemical detection of most M(hedtc)$_2$ or M(hedtc)$_3$ complexes. Finally, M(hedtc)$_2$ and M(hedtc)$_3$ complexes formed via the exchange reaction prior to chromatographic separation are also water soluble, so that a relatively high percentage of water with a correspondingly smaller quantity of organic solvent, such as methanol, can be used instead of the more expensive and more toxic water–acetonitrile mixtures used previously.

The alteration to the methodology as described in this section therefore achieves simplification of the experiment, a decrease in operational costs, improved performance and fewer problems with potentially hazardous and toxic organic solvents, in the automated continuous seven day operation of the METSCAN method for the on-line fully automated continual determination of metals in industrial effluents. The hedtc$^-$ ligand has been used previously for post-column complex formation [68], the separation of metal complexes following precolumn derivatisation [69] and for *in situ* and exchange chromatography [70]. Details on the stability of the zinc complex and the free ligand under a range of conditions are available from these studies.

The instrumentation used in the final version of the METSCAN system is closely related to that described previously. Figure 7.20 is a schematic diagram of the operational system. A reservoir of zinc bis(2-hydroxyethyl)-dithiocarbamate reagent is automatically mixed with the industrial effluent in a reactor coil. A bubble chamber removes any air bubbles formed during the mixing stage, and the sample is then automatically injected onto the chromatographic part of the system. The

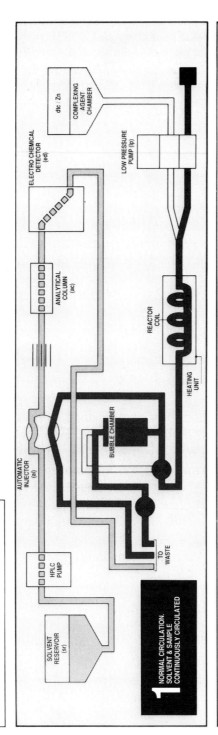

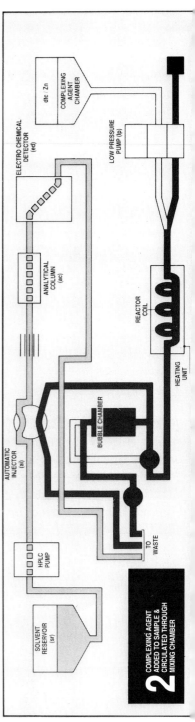

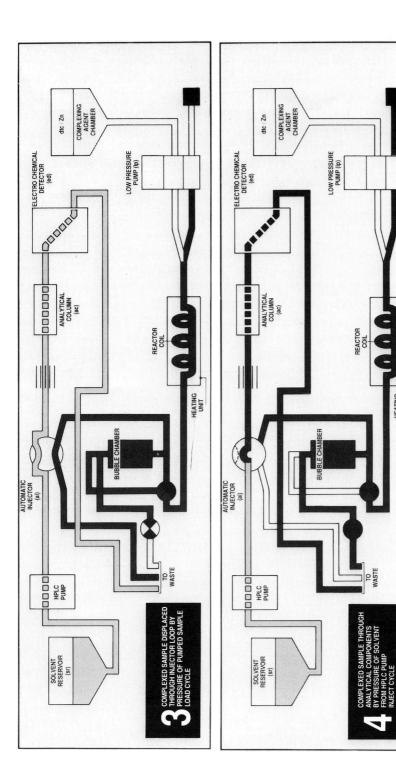

Fig. 7.20. Operation of the METSCAN instrument for the on-line determination of metals using the Zn(hedtc)₂ method.

flow rate of the eluting chromatographic solvent is carefully controlled in the 0.1 to 0.8 ml min^{-1} range. The glassy carbon working electrode, Ag/AgCl reference electrode, the potentiostat, chromatographic columns, as well as other aspects of the instrumentation, are the same as those described in the previous section. The chromatographic solvent was prepared by mixing an aqueous 0.025 M triethylammonium acetate buffer (pH 6.5) solution and methanol in the ratio 60:40. Methanol concentrations up to 80% can be used, but these decrease the retention times and increase the problem of organic solvent cost and toxicity.

The stability of Zn(hedtc)$_2$ was examined over a range of conditions. At a concentration of 1×10^{-4} M and in the presence of 0.025 M triethylammonium acetate buffer, no detectable change in Zn(hedtc)$_2$ concentration could be detected over a two week period. These zinc complex concentration and buffer conditions were therefore considered to be suitable for the METSCAN method, which requires a 1-week continuous operation period without replacement of reagents or operator intervention. As an alternative, a mixture of 1×10^{-4} M Zn(NO$_3$)$_2$ and 1×10^{-4} [NH$_4$][hedtc] can be used in the buffered solution. This has the advantage of ensuring that only a minimal amount of uncomplexed and readily oxidized free ligand hedtc$^-$ is present, because the equilibrium

$$Zn^{2+} + hedtc^- \rightleftharpoons Zn(hedtc)_2 \qquad (9)$$

is shifted further to the right hand side by the presence of excess zinc ions. The 10^{-4} M Zn(hedtc)$_2$ complex in aqueous buffered media is stored in the METSCAN instrument and, when required, is available for mixing with the industrial effluent via the use of a low pressure pump. The exchange reactions given in equations (7) and (8), enable M(hedtc)$_2$ and M(hedtc)$_3$ complexes to be formed after zinc reagent and industrial effluent are mixed. It is assumed that the metal ions of interest rapidly form the M(hedtc)$_2$ and M(hedtc)$_3$ complexes in the reaction chamber. If this is not the case, the heating unit can be used. In the industrial effluents examined, the metal ions are present at a concentration below 10^{-4} M, so there is a significant excess of Zn(hedtc)$_2$ present, to ensure that the equilibrium positions in equations (7) and (8) lie well to the right.

The complexes of the metals of interest formed by the exchange reaction, along with excess Zn(hedtc)$_2$, are automatically injected into the flowing chromatographic solvent. A potential of +0.9 V vs Ag/AgCl gives a very sensitive response for oxidation of nickel, copper, cobalt (metal based oxidation) and cadmium (ligand based oxidation) complexes and an insensitive response for Zn(hedtc)$_2$ (ligand based oxidation), so that this potential is very suitable for multi-element determinations. Measurement of the Zn(hedtc)$_2$ response is a useful marker to ensure that an excess of the reagent was actually present, and the decrease in peak height of this response relative to a control experiment provides information on how much hedtc$^-$ has been complexed by other metals present. If the Zn(hedtc)$_2$ response is not observed, then an insufficient concentration of this reagent is present, and calculated values of metal ions in the industrial effluents represent lower limits, rather than correct values.

In the absence of the triethylammonium acetate buffer in the chromatographic solvent, peak heights are non-reproducible. The buffer presumably blocks residual silanol groups on the C$_{18}$ column, which would otherwise be available for metal complexation [69]. The presence of this buffer in the zinc reagent also ensures that reproducible conditions are available for metal complex formation in the reaction chamber.

Figure 7.21A shows a chromatogram of the initially observed response for a mixture of Cu, Co, Cd and Ni metal ions in a standard solution of known composition. In the experiment described in Fig. 7.21, a new

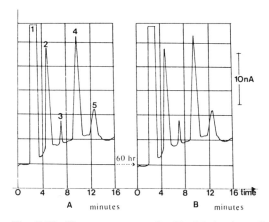

Fig. 7.21. Chromatogram of a 20 μl injection of a solution containing 500 ppb of Cu, Ni and Co and 1000 ppb of Cd using the Zn(hedtc)$_2$ method. A is the response observed initially and B the response after 60 hours of continual operation of the automated METSCAN system. Responses are due to: (1) excess of Zn(hedtc)$_2$; (2) Co(hedtc)$_3$; (3) Cd(hedtc)$_2$; (4) Ni(hedtc)$_2$; (5) Cu(hedtc)$_2$. Complexing reagent 10^{-4} M Zn(hedtc)$_2$. Altex C$_{18}$ Spherisorb Column. Chromatographic solvent, 40% methanol : 60% water (buffer). Flow rate, 0.8 ml min^{-1}. Electrode detection potential, +0.9 V vs Ag/AgCl; Temperature, 18°C. Reproduced by courtesy: *Anal. Chem.* 61 (1989) 1494.

References pp. 210–211

TABLE 7.3

VARIATION OF THE MEASURED CURRENT PEAK HEIGHT FOR COPPER, NICKEL, COBALT, AND CADMIUM AS A FUNCTION OF TIME OF CONTINUOUS OPERATION OF THE AUTOMATED METSCAN SYSTEM AND THE $Zn(hedtc)_2$ METHOD [a]

Time (hr)	Decrease in peak height with respect to the initial peak height (%)			
	Cu	Ni	Co	Cd
2	0.5	0.5	0.4	0.3
4	1.3	1.2	1.2	1.0
6	3.4	3.0	3.2	2.8
8	4.2	4.3	4.0	4.2
10	5.4	5.2	5.3	5.0
10 [b]	0.1	0.0	0.1	0.0

[a] Experimental conditions are given in the caption to Fig. 7.21. Data taken from reference [12].
[b] Data obtained after a cleaning potential of -1.2 V vs Ag/AgCl has been applied for 30 seconds.

injection of the same solution was undertaken every 20 minutes. The response in Fig. 7.21B, which was observed after 60 hours of running time, is virtually identical to the initially observed response in Fig. 7.21A.

After a 1 week continual operating period, with a sample injected every 20 minutes, less than 5% variation in peak height was found in data obtained via the automated method and the use of the exchange reaction with $Zn(hedtc)_2$, provided the glassy electrode was cleaned every 10 hours (or approximately every 60 injections), by application of a cleaning potential of -1.2 V vs Ag/AgCl for 30 seconds. In the absence of this cleaning pulse, a gradual deterioration in response is observed (peak diminishes in height). The application of the pulsed potential restores the response to the initial value. Data in Table 7.3 illustrate the gradual decay of the response and the restoration achieved by application of the pulse.

Figure 7.22 shows the chromatogram observed on drinking water which has passed through copper pipes and from river water. In these two examples, nickel and copper levels may be determined, but cobalt and cadmium levels are too low to be detected.

While the calibration curves are linear for each element, the sensitivity (peak current per unit concentration) is metal dependent. The limits of detection for each element are given in Table 7.4.

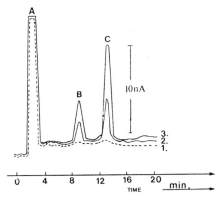

Fig. 7.22. Determination of copper and nickel by HPLCEC in: (*1*) distilled water; (*2*) river water; (*3*) drinking water. A = excess Zn(hedtc)$_2$; B = Ni(hedtc)$_2$; C = Cu(hedtc)$_2$. Experimental parameters and conditions as in Fig. 7.21. Reproduced by courtesy: *Anal. Chem.* 61 (1989) 1494.

TABLE 7.4

LIMITS OF DETECTION FOR THE DETERMINATION OF COPPER, NICKEL, COBALT AND CADMIUM BY THE FULLY AUTOMATED METSCAN SYSTEM AND THE Zn(hedtc)$_2$ METHOD [a]

Metal	Detection limit (ppb) [b]
Copper	20
Nickel	10
Cobalt	10
Cadmium	100

[a] Complex forming reagent, aqueous, buffered 1×10^{-4} M Zn(NO$_3$)$_2$/1×10^{-4} M [NH$_4$][hedtc]. Chromatographic solvent, 40% methanol/60% water (buffer). Flow rate, 0.8 ml min^{-1}. Detection potential $+0.9$ V vs Ag/AgCl. Altex C$_{18}$ Spherisorb Column. Temperature, 18°C. Data taken from reference [12].
[b] Calculated as a signal-to-noise ratio of 3 : 1.

Industrial liquors and effluents from electroplating or electrorefining industries often have considerably elevated cobalt and cadmium levels, which can be monitored routinely by the method described in this work for one week periods, without operator intervention or maintenance. The concept of using an exchange reaction with Zn(hedtc)$_2$, instead of *in situ* complex formation with an excess of dithiocarbamate ligand or *ex situ* complex formation in the reaction chamber, offers a considerable improvement in reproducibility, stability, cost of reagents and safety.

References pp. 210–211

This method has enabled the objective of long-term monitoring of metal ion concentrations to be achieved efficiently and effectively by chromatographic separation of metal complexes and electrochemical detection.

REFERENCES

1   G-A Junter (Editor), Electrochemical Detection Techniques in the Applied Biosciences, Volumes 1 and 2, Ellis Horwood, Chichester, 1988.
2   J. Wang, Electroanalytical Techniques in Clinical Chemistry and Laboratory Medicine, VCH Publishers, Weinheim, 1988.
3   A.M. Bond and G.G. Wallace, Anal. Chem., 53 (1981) 1209.
4   A.M. Bond and G.G. Wallace, Anal. Chem., 54 (1982) 1706.
5   A.M. Bond and G.G. Wallace, Anal. Chem., 55 (1983) 718.
6   Automated Metal Detection. A.M. Bond, G.G. Wallace and L. McLachlan. Australian Patent 3026/82: Filed 9th. March, 1982. European Patent Appl. EP88628 AZ: Filed 14th. September, 1983. See Chem. Abstr. 99 (1983) 186679a.
7   A.M. Bond and G.G. Wallace, J. Liquid Chromatog., 6 (1983) 1799.
8   A.M. Bond and G.G. Wallace, Anal. Chem., 56 (1984) 2085.
9   A.M. Bond and G.G. Wallace, Anal. Chim. Acta, 164 (1984) 223.
10 A.M. Bond, R.W. Knight, J.B. Reust, D.J. Tucker and G.G. Wallace, Anal. Chim. Acta, 182 (1986) 47.
11 A.M. Bond, W.N.C. Garrard, I.D. Heritage, T.P. Majewski, G.G. Wallace, M.J.P. McBurney, E.T. Crosher and L.S. McLachlan, Anal. Chem., 60 (1988) 1357.
12 A.M. Bond and T.P. Majewski, Anal. Chem., 61 (1989) 1494.
13 Y. Kitani, Bunseki Kiki Anal. Instrum., 14 (1976) 516.
14 P.C. Uden and I.E. Bigley, Anal. Chim. Acta, 94 (1977) 29.
15 G. Schwedt, Chromatographia, 12 (1979) 613.
16 G. Schwedt, Top. Curr. Chem., 85 (1979) 159.
17 H. Veening and B.R. Willeford, Rev. Inorg. Chem., 1 (1979) 281.
18 K. Saitoh, Bunseki Chem. (1979) 548.
19 G. Schwedt, Chromatographische Methoden in der Anorganischen Analytik, Huethig, Heidelberg (1980).
20 A.R. Timerbaev, O.M. Petrukhin and Yu. A. Zolotov, Zh. Anal. Khim., 36 (1981) 1160.
21 B.R. Willeford and H. Veening, J. Chromatogr., 251 (1982) 61.
22 H. Veening and B.R. Willeford, Adv. Chromatogr., 22 (1983) 117.
23 J. Maslowska and G. Bazylak, Zesz. Nauk-Politech, Lodz., Chem. Spoz, (1983) 91. Chem. Abstr. 101 (1984) 117239e.
24 J.W. O'Laughlin, J. Liquid Chromatogr., 7 (1984) 127.
25 I.S. Krull, in Liquid Chromatography in Environmental Analysis. Humana Press, Clifton, 1984, p. 169.
26 G. Nickless, J. Chromatogr., 313 (1985) 129.
27 A.R. Timerbaev, O.M. Petrukhin and Yu. A. Zolotov, Fresenius Z. Anal. Chem., 327 (1987) 87.
28 G. Schwedt, Chromatographia, 11 (1978) 145.
29 R.M. Smith and L.E. Yankey, Analyst (London), 107 (1985) 744.

30   R.M. Smith, A.M. Butt and A. Thakur, Analyst (London), 110 (1982) 35.
31   S. Ichinoki, T. Morita and M. Yamazaki, J. Liq. Chromatogr., 6 (1983) 2079.
32   S. Ichinoki, T. Morita and M. Yamazaki, J. Liq. Chromatogr., 7 (1984) 2467.
33   S. Ichinoki and M. Yamazaki, Anal. Chem. 57 (1985) 2219.
34   B.J. Mueller and R.J. Lovett, Anal. Chem., 57 (1985) 2693.
35   B.J. Mueller and R.J. Lovett, Anal. Chem., 59 (1987) 1405.
36   J.N. King and J.S. Fritz, Anal. Chem., 59 (1987) 703.
37   H. Ge and G.G. Wallace, Anal. Chem., 60 (1988) 830.
38   J.N. King and J.S. Fritz, Anal. Chem., 57 (1985) 1016.
39   Y-T. Shih and P.W. Carr, Anal. Chim. Acta, 142 (1982) 55.
40   A. Berthod, M. Kolosky, J.L. Rocca and O. Vittori, Analusis, 7 (1979) 395.
41   H. Hoshino and T. Yotsuyanagi, Bunseki Kagaku 29, (1980) 807.
42   A.M. Bond and Y. Nagaosa, Anal. Chim. Acta, 178 (1985) 197.
43   C. Baiocchi, G. Saini, P. Bertolo, G.P. Cartoni, and G. Pettiti, Analyst, 113 (1988) 805.
44   R.C. Gurira and P.W. Carr, J. Chromatogr. Sci., 20 (1982) 461.
45   M. Tabata and M. Tanaka, Anal. Lett., 13 (1980) 427.
46   D.A. Roston, Anal. Chem., 56 (1984) 241.
47   H. Hoshino, T. Yotsuyanagi and K. Aomura, Bunseki Kagaku 27 (1978) 315.
48   H. Hoshino and T. Yotsuyanagi, Talanta, 31 (1984) 525.
49   H. Hoshino and T. Yotsuyanagi, Anal. Chem., 57 (1985) 625.
50   J.B. Noffishinger and N.D. Danielson, J. Liq. Chromatogr., 9 (1986) 2165.
51   X.S. Zhang, X.P. Zhu and C.S. Lin, Talanta, 33 (1986) 838.
52   J.E. DiNunzio, R.W. Yost and E.K. Hutchison, Talanta, 32 (1985) 803.
53   P.T. Kissinger and W.R. Heineman (Editors), Laboratory Techniques in Electroanalytical Chemistry, Marcel Dekker, New York, 1984.
54   T.H. Ryan (Editor), Electrochemical Detectors, Plenum Press, New York, 1984.
55   M.R. Smyth and J.G. Vos (Editors), Electrochemistry, Sensors and Analysis, Elsevier, Amsterdam, 1986.
56   D.C. Johnson, S.G. Weber, A.M. Bond, R.M. Wightman, R.E. Shoup and I.S. Krull, Anal. Chim. Acta, 180 (1986) 187.
57   D. Coucouvanis, Prog. Inorg. Chem., 11 (1970) 233, 26 (1979) 301.
58   A.M. Bond and R.L. Martin, Coord. Chem. Rev., 54 (1984) 23.
59   R.J. Magee, Rev. Anal. Chem. 1 (1973) 335.
60   A. Hulanicki, Talanta, 14 (1967) 1371.
61   S.R. Koistyohann and J.W. Wen, Anal. Chem., 45 (1973) 1986.
62   G.K. Budnikov, V.F. Toropova, N.A. Ulakhovich and I.P. Vites, Zh. Anal. Chem., 30 (1975) 2120.
63   H. Kitamura, A. Ichimura and T. Kitigawa, Nippon Kagaku Kaishi, (1979) 354.
64   A.R. Hendrickson, R.L. Martin, and N.M. Rohde, Inorg. Chem. 15 (1976) 2115.
65   G. Cauquis and D. Lachenal, J. Electroanal. Chem., 43 (1973) 205.
66   C. Scrimage and L.J. Dehayes, Inorg. Nucl. Chem. Lett. 14 (1978) 125.
67   A.M. Bond, Modern Polarographic Methods in Analytical Chemistry, Marcel Dekker, New York, 1980.
68   J.B. Flato, Anal. Chem. 44 (1972) 75A.
69   P. Maitoza and D.C. Johnson, Anal. Chim. Acta, 118 (1980) 233.

I.S. Krull (Ed.), *Trace Metal Analysis and Speciation*
Journal of Chromatography Library Series, Vol. 47
© 1991 Elsevier Science Publishers B.V., Amsterdam

Chapter 8

Ion Chromatographic Speciation of Trace Metals

D.T. GJERDE[1] and H.C. MEHRA[2]
[1] *Sarasep, Inc., 1600 Wyatt Dr., Suite 10, Santa Clara, CA 95054, U.S.A.*
[2] *Chemical Waste Management, Inc., P.O. Box 4249, Modesto, CA 95352, U.S.A.*

8.1. INTRODUCTION

The term chemical speciation is the identification and determination of the individual amounts of the various chemical forms of an element.

Several elements such as As and Se are converted to different species during weathering, uptake by biota, elimination from biota, fixation in sediments, or remobilization. In order for an analysis to be successful the metal species must be stable throughout the entire analysis.

Handling of the sample may result in changes in the amounts or types of species. The chromatography or sample preparation may change the sample. This may prevent the accurate measurement of some metal species. In fact, the literature may be slanted toward those species that are relatively easy to do. For example, there are several papers in the literature on the separation of Fe(II) and Fe(III), but there are very few reports on the separation of V(IV) and V(V). Thus, there are many opportunities for future work.

There are several reasons for performing speciation. Of course there are the academic studies of various types of trace metals and their conversion and equilibria of various forms. Industrial speciation is important. The effectiveness of chromium plating baths depends on the amounts of Cr(III) and Cr(VI) that are present. Gold plating processes involve the use of Au(I) and Au(III) cyanide complexes. Vanadium speciation is important in the recovery of sulfur from geothermal water.

But environmental applications appear to be of greatest current interest. Speciation is important because each inorganic species of a particular element may possess quite different biological, medicinal and toxicological properties. There are differences of element-specific species in physicochemical reactivity, biological availability and element transport in the environment and into the food chain. Knowing the chemical forms of an element in environmental, agricultural, or other samples can be much more important than knowing the total elemental content. For example, pollution incidents involving mercury have shown that total metal data are insufficient and often misleading in assessing the potential hazard of this metal. The general application of ion chromatography (IC) to environmental samples has been described in a recent review [81]. The first part of this chapter describes the IC analytical tools needed for speciation. Detection methods are described in greater detail elsewhere; only a short discussion is given in this chapter. Factors affecting the ion-exchange chromatography used in many of these separations are discussed. Many of these factors are also important in ion-pairing chromatography. The second part of this chapter lists several of the papers describing speciation and the conditions needed to perform the analysis.

8.2. ANALYTICAL TOOLS

Perhaps the greatest limitation in understanding the environmental role of chemical species has been the lack of suitable chemical methods and procedures to identify and quantify different species at the concentrations found in the environment [1]. Often species have to be preconcentrated to reach concentrations sufficiently high for detection. But preconcentration can transform or shift the equilibria of the various matrix species. Gas chromatography [2] and high performance liquid chromatography (HPLC) [2–4] have been used to a limited extent. However, high gas chromatography temperatures may cause decomposition. Thus a sensitive, nonobtrusive method is needed. Until the introduction of ion chromatography, HPLC studies have been limited to ion-pairing reagents, IC is now part of HPLC with an emphasis on (but not exclusive to) ion-exchange separation. There are literally hundreds of references describing the separation of various elements of specific oxidation states. There exist significantly fewer references that describe the simultaneous separation of different species of a particular element. Papers describing the speciation of an element in a single run are emphasized in this chapter. Tables describing the chromatographic conditions are listed for each element species. Because capacity factor data were not generally available, retention times are given for each species for comparison purposes only.

8.2.1. Detection

The type of detection employed in ion chromatography is important to the type of chromatography needed. Detectors that are sensitive or can be "tuned" for an element are the most useful for speciation. For example, it was shown in a review by Urasa [5] on DC plasma atomic emission spectrometry (DCPAE) detection that only one of the metal species need be retained by the chromatographic column in order to perform a speciation analysis. Cr(III) exists as a cation in aqueous solution. Cr(VI) exists as an anion i.e. CrO_4^{2-}. If one is using a cation exchange column and injects a mixture of Cr(III) and Cr(VI), then the anion, Cr(VI) elutes with the void volume. Selective detection allows quantification of this peak in spite of the fact that many other materials are eluting at the same time. Cr(III) is eluted later as part of the cation exchange process and detected. The eluant conditions needed to elute this peak are not very stringent. All that is needed is to elute Cr(III) quickly without changing its form. Since the eluant is not detected, the eluant type, pH, and/or concentration can be changed as needed. Thus, methods for speciation can be developed quickly and performed easily provided that there is access to selective detection.

But selective detection is not always available. There have been several reports where general detection methods such as conductivity or post-column reaction photometric detection have been used. The chromatography is much more important in these cases. With general detection, the metal species must be the same ionic charge. Non-retained sample peaks that elute with the void volume generally contain other components that cause a detector response. This would interfere with quantification of the non-retained peak.

Strong chelating or complexing agents may form anionic complexes with metals. These anionic complexes can be separated by anion-exchange chromatography. Continuing with the chromium example, Cr(III) can be converted to an anion complex by complexation with EDTA. Thus, a mixture of $CrEDTA^-$ and CrO_4^{2-} may be separated by anion-exchange chromatography and detected by UV detection.

8.2.2. Chromatography

The majority of speciation separations are performed by ion exchange chromatography although there has been significant work performed by ion-pair chromatography. Many of the chromatographic discussions in this chapter refer specifically to ion-exchange chromatography although many of the concepts also may be applied to ion-pair chromatography.

The ion-exchange separation of metal species in a chromatographic system can be performed based on either of two concepts. Sample species can be separated on a column based on affinity differences of the species for the column. Sample species may also be separated by using a complexing reagent in the eluant. The complexing reagent changes the form of the sample species allowing it to move down the column more easily.

These concepts are sometimes called the "push/pull" mechanism of ion-exchange chromatography [6]. The eluant that competes with sample species for ion-exchange sites and elutes the sample from the column is operating under a "push" mechanism. The eluant "pushes" or elutes the samples species from the column.

Adding a complexing reagent to the eluant and converting the sample species into a complex that is not as well retained by the ion-exchange column is called the "pull" mechanism. The complexing agent "pulls" the sample species down the column.

The effect can be shown by equations. It has been shown that the ion-exchange reaction of the eluant and sample species competing for the ion exchangers can be written as an equilibrium equation [7]. Using chromatographic theory, the equilibrium equation can be rearranged to predict behavior in ion-exchange chromatography. The result is shown by the equation:

$$\log k = a/e \, \log C - a/e \, \log E + 1/e \, \log K_{eq} - a/e \, \log e + D$$

where k is the sample capacity factor, a is the charge of the sample ion, e is the charge of the eluant driving ion, C is the resin capacity, E is the eluant concentration, and K_{eq} is the static solution ion-exchange equilibrium constant. The constant D includes the terms column void volume and resin density.

Examination of the equation shows that controlling separations (changing $\log k$ or retention) by affinity differences is not very powerful. Options for controlling the separation are limited. The resin capacity, sample ion charge, and affinity of a sample (K_{eq}) for a particular column is usually constant or not easily changed. The only real variables are the eluant type (affinity of eluant ion for the ion exchanger), eluant ionic charge, and the eluant pH and concentration. Some retention crossovers can be achieved by using different eluant types and varying the eluant concentration.

But the real power for performing ion chromatographic separations occurs when a complexing agent is added to the eluant. The complexing agent controls the amount of ion species available to compete with the eluant for the ion exchanger. In effect, the K_{eq} term is changed (usually lowered). In cation chromatography, K_{eq} (and log k) is usually lowered because neutral or anionic complexes are formed. In anion chromatography, K_{eq} and retention is usually increased. The complexing agent is usually specific for a metal or group of metals. Thus, a particular separation can be achieved by choosing the type, concentration and pH of the complexing agent added to the eluant. There are no hard and fast rules on choosing a particular complexing agent to perform a particular separation. Probably, by going through the literature and making correlations, a paper could be written on the subject, but this has not yet been done.

8.3. SPECIATION OF METALS

8.3.1. Chromium

Chromium is known in all oxidation state between 0 and VI, but is commonly found in oxidation states III and VI. Cr(III) species in acidic solution exist as $Cr(H_2O)_6^{3+}$ ions and in concentrated alkali have been identified as $Cr(OH)_6^{3-}$ and $Cr(OH)_5(H_2O)^{2-}$. These ions are regular octahedral. Cr(VI) in basic solution above pH 6 exists as tetrahedral yellow chromate, CrO_4^{2-} ion. Between pH 2 and 6, $HCrO_4^-$ and the orange-red dichromate, $Cr_2O_7^{2-}$, are in equilibrium. Under strongly acid conditions, only dichromate ion exists. Addition of alkali to dichromate gives chromate.

Chromium metal is produced by roasting the chromate ore. The largest single use of the metal is as corrosion inhibitors in alloying steels. Chromium finds its way into the environment through industrial wastes from electroplating sludge, tannery wastes, the manufacture of corrosion inhibitors, and municipal sewage sludge. These sludges are of particular concern. Cr(III) is essential to human nutrition. Cr(III) is present in most soils while Cr(VI) is only occasionally present. Cr(III) is less toxic and less mobile (in the environment) than Cr(VI). Cr(VI) compounds are very toxic to aquatic plants and animal life as evidenced by their widespread use as algaecides. Cr(VI) toxicity may manifest itself in the form of skin ulceration, nasal perforation, and lung cancer.

There have been several reports [8–13] describing the speciation of chromium. Table 8.1 describes the conditions used in these reports. Many of these reports described using DC plasma atomic emission (DCPAE), ICP atomic emission (ICPAE) and atomic absorption (AA)

TABLE 8.1
SPECIATION OF CHROMIUM

Sample	Species (ret. time, min)	Eluant	Column	Detector	Detection limits	Ref.
Wastewater	Cr(III) (0.9), Cr(VI) (4.0)	0.5 mM phthalate pH 6.5	TSKgel IC Anion PW	ICPAE	7–22 ng	9
Plating bath	Cr(III) (3.5), Cr(VI) (4.8)	2mM PDCA, 2 mM Na$_2$HPO$_4$ 10 mM NaI, 50 mM NH$_4$Ac, 2.8 mM LiOH	Dionex HPIC CS5	UV, 520 nm after PCR rxn	0.3–30 ppb	10
Solution	Cr(VI) (12), Cr(III) (17)	0.375 mM EDTA, pH 4.6	2 Vydac 302 IC in series	UV, 350, 220 nm	2 ppm	11
Solution	Cr(III) (1.4), Cr(VI) (4.6)	50 mM HNO,	Dionex AS7	DCPAE	1 ppb	13
Solution	Cr(VI) (1.9), Cr(III) (5.2)	(a) 7.5 mM citrate, 10 mM oxalic acid, then inject 1 M HCl (b) Water with multiple injections of 1 M HCl	Dionex CS2	DCPAE	1 ppb	13
Solution	Cr(VI) (1.9), Cr(III) (5.2) Cr(III) (0.9), Cr(VI) (8.2)	(a) 5 mM pentane n-sulfonate (b) 5 mM tetrabutyl ammonium salt, pH 2.9	Ultrasphere ODS	ICPAE	3–15 ppb	14

selective-type detection. UV/VIS detection has also been extensively used. Atomic emission detection has been found to be very promising due to its selectivity and low detection limits.

The basis of many of the separations has been to convert Cr(III) to an anion by adding a complexing agent. Cr(VI) is already an anion (usually CrO_4^{2-}), hence, anion chromatography can be used to separate mixtures. An example is separations carried out in the presence of KSCN converting Cr^{3+} to $Cr(SCN)_4^-$.

Pyridine dicarboxylic acid (PDCA) was used as a complexing agent for Cr^{3+} to form anionic $Cr(PDCA)_2^-$. The sample pH is critical to the separation. Optimum results were reported at pH 6.8. Both Cr(III) and Cr(VI) were detected using UV/VIS detection and diphenylcarbohydrazide as a post-column reagent. The method was found to be applicable for plating baths and waste water analysis.

Geddes and Tarter [11] reported the use of EDTA eluant and UV detection for Cr speciation. Two silica-based anion exchange columns were connected in series. However, the chromatographic peaks were very broad and the method had poor detection limits.

The simultaneous determination of Cr(III) and Cr(VI) in industrial effluent using AA detection was evaluated [12]. The method employed a "mixed-moded" anion-exchange/cation-exchange column. An acid eluant eluted Cr^{3+} first. This was followed by a carbonate/bicarbonate step gradient to elute Cr(VI) anion. The flow rate was unusually high (7 ml min^{-1}) resulting in a 1 minute separation time for both species. Interferences were minimal due to the element specific nature of the atomic absorption as detection system. The presence of 1000 ppm of SO_4^{2-} did not affect the results for industrial effluent samples. Recently, Urasa et al. [13] developed a method for Cr speciation using both anion and cation separator columns. The procedure used DCPAE detection and required a preconcentration step to achieve a detection limit of 1 ppb. The method was applied to human serum, natural waters and industrial stream samples. Krull et al. [14] developed two ion-pairing methods and ICPAE detection. Both pentane n-sulfonate and tetrabutylammonium ion pairing reagents were used. Depending on the type of ion pairing reagent, charge of the species, and type of ion exchangers used, either Cr(III) or Cr(VI) will elute with the void and the other species will be retained. The detection limits were quite good at the 3–15 ppb level.

8.3.2. Iron

Iron is the second most abundant metal after aluminum. The highest oxidation state of iron is VI, although of course, II and III oxidation states are the most common. Ferrous iron forms a variety of complexes. In aqueous solution Fe(II) exists as $Fe(H_2O)_6^{2+}$, which is pale sea-green in color. Fe(II) is slowly oxidized in acid. Most Fe(II) complexes are octahedral. Perhaps, the most important ferrous iron complex is haem which exists in hemoglobin.

In aqueous solution, Fe(III) shows a strong tendency to hydrolyze. The hydrated ion complex $Fe(H_2O)_6^{3+}$ is pale purple. It exists only in strongly acid solutions at about pH 0. In less acid media, hydroxy complexes are formed. In the presence of complexing anion such as Cl^-, the hydrolysis of Fe^{3+} or of $FeCl_3$, is more complicated giving chloro, aqua, and hydroxy species. Fe(III) complexes are mainly octahedral just as they are for Fe(II). Fe(III) has a greater affinity for oxy-ligands whereas Fe(II) has a slight preference for ligands containing nitrogen donor atoms.

The simultaneous determination of Fe(II) and Fe(III) is important to understanding the environmental redox processes in biological systems. Iron activity affects several chemical processes in natural waters and its speciation concentration is a significant factor in the evaluation of water quality.

Iron speciation of Fe(II) and Fe(III) is reported more often than any other speciation. The methods are based on cation-exchange, anion-exchange, and ion-pairing chromatography. The reader is invited to study Table 8.2 for details. Only a few of the methods are discussed here.

Saitoh and Oikawa [15] simultaneously determined Fe(II) and Fe(III) using post-column reaction (PCR) detection. The PCR reagent was bathophenanthrolinedisulfonic acid and ascorbic acid. This procedure was found to be successful in spring water samples. Moses et al. [16] reported the determination of Fe(II) and Fe(III) in water samples. The detection system consisted of PCR reaction with PAR detection reagent. In this work, the Fe(II)/Fe(III) ratio increased at times—probably due to Fe(III) being photochemically reduced.

The presence of trace iron contaminants in gold plating baths can cause brittle deposits. $Fe(CN)_6^{4-}$ and $Fe(CN)_6^{3-}$ were reported to be separated by anion chromatography [17]. These complexes are multivalent and difficult to elute. Small amounts of Na_2CO_3 added to the mobile phase sharpened the peaks.

8.3.3. Arsenic

Arsenic is found in igneous and sedimentary rocks. The most common commercial source is as the by-product from the refining of copper, lead, cobalt, and gold ores. Although arsenic is actually a metalloid, it is grouped with metals for most environmental purposes.

Arsenic chemistry is complex, involving a variety of oxidation states, both as anionic and cationic species, and both inorganic and organic compounds. Of these, (III), (V), and (−III) are the most common oxidation states. To further complicate matters, the oxidation states of arsenic change easily and reversibly. As(III) is commonly encountered as the arsenite ion, $H_2AsO_3^-$. Arsenious acid is a weak acid, $pK_{a1} = 9.2$, $pK_{a2} = 13$.

As(V) exists as $H_2AsO_4^-$ in aqueous solution. Arsenic acid is a weak tribasic acid. Its dissociation constants, $pK_a = 2.3$, 6.8, and 11.5, are similar to phosphoric acid. Oxidation of As(III) to As(V) in dissolved oxygen is slow at neutral pH, but is much faster at either extreme [35]. In reducing environments, As(III) is produced, but As(V) is the most stable state in aerobic environments.

Most of the arsenic compounds are highly toxic causing dermatitis, acute and chronic poisoning, and possibly cancer. Arsenic is found in virtually all soil and other environmental matrices [36]. Arsenic is present in coal, pesticides, preservatives, etc. Arsenite, a commercial form of arsenic, is one of the most toxic forms of arsenic.

Arsenic speciation by IC can be simple and reproducible (Table 8.3). One procedure, using a suppressed system was reported by Takamatsu et al. [37]. The method required two separate injections for the determination of AsO_3^{3-} and AsO_4^{3-}. The procedure was sensitive and there was no interference by ions such as NO_3^-, HPO_4^{2-}, and SO_4^{2-} that are present in the sediment extract.

Ricci et al. [38] used gradient elution separation for the determination of five organic and inorganic species of arsenic. Although the method is effective and separates all five species in one run, the column must be restabilized with the weaker eluant for 1 hour after each run. For this reason, the authors suggested using two separate isocratic approaches with two eluant buffer systems. The method was applied for the determination of organoarsenic compounds in air, but could be used for other environmental samples.

Ion-exclusion chromatography [39] was used to separate AsO_3^{3-} and AsO_4^{3-}. UV detection was used to measure both species although the sensitivity for arsenate was not good. Anion exclusion was also

TABLE 8.2
SPECIATION OF IRON

Sample	Species (ret. time, min)	Eluant	Column	Detector	Detection limits	Ref.
Solution	Fe(III) (2), Fe(II) (4)	15 mM oxalic acid, or 25 mM citric acid, or 150 mM lactic acid, or 50 mM tartaric acid, pH 4.5	TSK-gel IC Cation	UV, 530 nm after PCR rxn with batho-phenanthroline/ ascorbic acid	10 ppb	15
Transition metals	Fe(III) (4.6), Fe(II) (12.5)	6 mM 2,6-pyridinedi-carboxylic acid, 100 mM acetate, pH 4.5	Dionex CS5	UV, 520 nm after PCR rxn with PAR		16
Transition metals	Fe(III) (3.2), Fe(II) (18), and	5 mM oxalic acid, 3.75 nM citric acid, pH 4.37 with LiOH	Dionex AS7 and 2 CS5	conductivity	1–10 ppm	18
Transition metals	Fe(III), (3), Fe(II) (20)	120 mM lactate or 12 mM HIBA or tartrate gradient	Vydac 401 SA	UV, 540 mm after PCR rxn with PAR	2–20 ng	19
Lanthanide, other metal mixtures	Fe(III) (5), Fe(II) (14)	2,6-pyridinedicarboxylic acid eluant, gradient and isocratic	Dionex CS5	UV, 520 nm after PCR rxn with PAR	20–40 ppb	20
Transition metals	Fe(III) (7.5), Fe(II) (20)	60 mM HCl/acetone to 60 mM HCl/acetone/DMF gradient	Amberlite SCX	UV after PCR rxn with PAR	10 nM	21
Transition metals	Fe(III) (1.9), Fe(II) (14),	10 mM citrate, 7.5 mM tartrate, pH 4.3, others	Dionex CS2 others	UV, 520 nm after PCR rxn with PAR	50 ppb	22, 24–25
Transition metals	Fe(III) (2.5), Fe(II) (8.3)	lactic acid	Partisil 10 SCX	UV 610 nm after PCR rxn with erio-chrome black T	2–10 ng	23

Sample	Species (conc.)	Eluent	Column	Detection	Detection limit	Ref.
Transition metals	Fe(III) (2), Fe(II) (5.7)	5 mM tartrate, 7.5 mM ethylenediamine, pH 4.5	Nucleosil 10-SA	Cu wire potentiometric	10 nmole	26
Human serum	Fe(II) (7), Fe(III) (21)	6 mM 2,6-pyridinedicarboxylic acid, pH 4.8 with LiOH	Dionex CS2	UV, 520 nm after PCR rxn with PAR	50 ppb	27
Transition metals	Fe(III) (3.5), Fe(II) (7.2)	4 mM 2,6-pyridinedicarboxylic acid, 2 mM Na_2SO_4, 15 mM NaCl	Dionex CS5	indirect amperometric after PCR rxn with Zn/DTC	1 ppm	28
Ores	Fe(III) (7), Fe(II) (31)	0.1 tartrate, 0.12% NaCl, pH 2.25	Nucleosil 10-SA	UV 495 nm after PCR rxn with PAR/ZnEDTA	10–20 ppb	30
Lanthanides	Fe(III) (5), Fe(II) (15)	9% diglycolic acid	Dionex Anion	UV 520 nm after PCR rxn with PAR	20–40 ppb	31
Plating bath cyanide complexes	Fe(III) Fe(II)	(a) 6 mM 2,6-pyridine-dicarboxylic acid, 8.6 mM LiOH	Dionex CS5	UV	–	17
		(b) 2 mM tetrabutyl-ammonium hydroxide, 0.2 mM Na_2CO_3, 35% ACN	Dionex MPIC NS1	suppressed conductivity		
Gold processing solutions, cyanide complexes	Fe(II) (10.8), Fe(III) (17.8)	5 mM tetramethyl-ammonium hydroxide, 23% ACN	Waters Nova-Pak C18	UV, 214 nm	1–3 ng	32
Transition metals	Fe(III) (3.2), Fe(II) (15)	2 mM octanesulfonate, 50 mM tartaric acid, pH 3.4	Waters μBondapak C18	UV 520 nm after PCR rxn with PAR	10 ppb	33
Solution	Fe(III) (4.3), Fe(II) (17)	50 mM Na dodecylsulfate, 1 mM sulfuric acid, 2.5 mM dodecanol, 2% ethylacetate	Rainin Microsorb ODS	UV after PCR rxn with PAR	50 ppb	34

TABLE 8.3

SPECIATION OF ARSENIC

Sample	Species (ret. time, min)	Eluant	Column	Detector	Detection limits	Ref.
Sediment extract	As(III), As(V)	carbonate/bicarbonate	Dionex Anion	suppressed conductivity		37
Environmental	dimethylarsenate (2), As(III) (3), monomethylarsenate (5), p-aminophenylarsenate (6), As(V) (8)	carbonate/borate gradient	Dionex Anion	AA	10 ppb	38
Solution	As(III) (10.9), As(V) (12.2)	3 mM $NaHCO_3$, 2.4 mM Na_2CO_3	Dionex Anion	UV after suppressor		39
	As(V) (9), As(III) (15)	10 mM HCl	Dionex Anion Exclusion	UV after suppressor		
Solution	As(V) (10), As(III) (16)	5 mM H_2SO_4	Wescan Anion Exclusion	nonsuppressed cond. and amperometric		40
Metallurgical media	As(III) (0.85), As(V) (4.05)	2.8 mM $NaHCO_3$, 2.25 mM Na_2CO_3	Dionex AS4	suppressed cond., pulsed amperometric and UV	0.3 ppb	41
Solution	As(III) (2.0), As(V) (6.0)	3 mM $NaHCO_3$, 2.4 mM Na_2CO_3	Dionex AS4	DCPAE	2.5 ppb	41
Solution	As(III) (5.0), As(V) (8.0)	170 mM $NaNO_3$	YSG Anion	coulometric	0.5 ppm	43
Mineral water	As(V) (9), As(III) (14)	10 mM orthophosphoric acid	Aminex HPX-87H	amperometric	12 nM	44
Solution	As(V), As(III)	0.8 mM KHP, 20 mM Tris, 24% methanol	Hamilton PRP-1 coated with centrimide	UV		45

Sample	Species (retention times)	Eluent	Column	Detection	Detection limit	Ref.
River water	dimethylarsinate (2.5), methylarsonate (5.0), p-aminophenylarsonate (9.8), phenylarsonate (9.8), o-aminophenylarsonate (9.8), As(V) (25)	2 mM NaNO$_2$	TSKgel IC-Anion-PW	conductivity	0.3–5.0 ppm	46
Urine	dimethylarsine (2.3), monomethylarsine (2.8), As(III) (4.2), arsenobetaine (5.0), As(V) (8.1)	3 mM NH$_4$H$_2$PO$_4$, pH 6	Hamilton PRP-X100, Vydac 201Tp	ICPAE	0.5 ppm	47
Urine, seawater	As(III) (1.4), dimethylarsine (2.0), monomethylarsine (2.8), As(V) (7.5)	10 mM (NH$_4$)$_2$CO$_3$, 2.5 mM Na$_2$SO$_4$, pH 8.2	Hamilton PRP-X100	ICPAE	0.5 ppm	48
Shale oil fraction	As(III) (5.2), monoethylarsonite (6.3), dimethylarsonate (6.8), As(V) (13.7), benzenearsonate (19)	5 mM tetrabutyl-ammonium phosphate, 10% methanol	Partisil ODS-3	ICPAE	40 ppm	49
Urine	As(III) (2.5), monomethylarsonic acid (4.5), dimethylarsonic acid (7.0), As(V) (8.9)	30 mM phosphate, pH 6.2	Ionspher SAX	hydride AA	2 μg	50
Metallurgical media	As(III) (0.88), As(V) (4.2)	2.8 mM NaHCO$_3$,	Dionex AS4	suppressed cond., amperometric	2–5 ppb	51
Solution	As(III) (2.0), dimethylarsenate (3.0), As(V) 8.8	5 mM tetrabutyl-ammonium	Alltech RP-18	ICPAE	1 ppm	52
Solution	As(III), As(V), arsenobetaine, arsenocholine	water/ACN/acetic acid 5 mM Na dodecyl-benzene-sulfonate	C18	GFAA	1 μg	53
Solution	methylarsonate (17), dimethylarsinate (40), arsanilic acid (49)	50 mM NaH$_2$PO$_4$	SAX	GPAA	10 ppb	54

used by Jupille [40] and amperometric detection employed for As(III).

IC was applied to the determination of AsO_3^{3-} and AsO_4^{3-} in metallurgical media [41]. Two types of procedures were proposed. One was using tripolar pulsed amperometric detection for As(III) and conductivity detection for As(V). The other procedure proposed using UV detection. UV was found to be faster and simpler, but less sensitive than electrochemical or conductivity detection. The effects of HCl and H_2SO_4 acids were discussed in this work. The method required a cleanup step to remove interferences or else partial precipitation of arsenic due to cations could result. Sample matrix can pose serious problems for conductivity and UV methods of detection.

The capabilities of direct current plasma atomic emission spectrometry as a powerful IC detection system for As(III) and As(V) was evaluated by Urasa and Ferede [42]. The presence of several common anions was studied. One fundamental drawback of DCP in this particular work appeared to be low sensitivity. But the use of a large sample loop or concentrator column provided improvement.

Coulometric detection was used by Hu and Tang [43] for AsO_3^{3-} and AsO_4^{3-} speciation. A nitrate eluant was used. The proposed method may be used in industrial and environmental samples.

Wet chemistry sample preparation was employed in an ion exclusion method by Butler [44]. As(III) and then, in a separate run, total arsenic were determined using electrochemical detection. As(V) was reduced by SO_2 to As(III) before quantification. A phosphoric acid eluant was found to be superior over HCl because it is less corrosive.

A modified nonpolar stationary phase and UV detection was used for the separation of arsenic species [45]. Unfortunately, little real sample work was performed. Recently, a nonsuppressed IC method for the separation of various organic arsenic anions was reported by Hirayama and Kuwamoto [46].

8.3.4. Tellurium

Tellurium is used in the metallurgical industry as an alloy constituent. Tellurium improves the acid resistance of lead used in batteries. It is also used in the manufacture of heat and abrasive-resistant rubber. Tellurium is obtained as a by-product in the electrolytic refining of copper. In the semiconductor industry, the ultra-trace level determination of tellurium in tellurium-doped single crystals is often required. Tellurium species are highly toxic.

Tellurium and selenium resemble each other chemically. The analytical chemistry of both elements is usually presented together. Tellurium and selenium are commonly found in oxidation states $(-II)$, (IV), and (VI) as well as in the elemental form. Oxidation state (IV) is the most stable.

In aqueous chemistry, tellurium is mainly found as telluride (Te^{2-}), tellurite (TeO_3^{2-}), and tellurate (TeO_4^{2-}). Zolotov et al. [55] developed an IC procedure for the separation of TeO_3^{2-} and TeO_4^{2-} using a suppressed ion chromatographic system. In the method, F^- interfered with the determination of TeO_3^{2-}, while SO_4^{2-} interfered with the determination of TeO_4^{2-}. In a report by Chen et al. [56], a similar method was used to speciate tellurium. The detection limits for TeO_4^{2-} were very poor in both methods. The studies were confined only to standard solutions.

8.3.5. Selenium

Selenium is widely dispersed. It is found, for example, in igneous rocks, volcanic sulfur deposits, hydrothermal deposits, and copper ores. Selenium is used in the electronics industry for the manufacture of rectifiers and photoconductivity cells. Selenium and its compounds are also used as additives in chromium plating, glass, ceramics, pigment, rubber, photography, lubricants, pharmaceuticals, and organic substances.

Selenium is both a toxic and an essential element. The toxicity depends greatly on the species. Selenium is a cumulative toxic substance and can be a serious health hazard when present in high concentrations in food and water. However, at very low levels ($\mu g\ kg^{-1}$), it is recognized as an essential trace element in animal nutrition.

There has been an increasing interest in the determination of selenium at trace levels in a wide variety of matrices. Selenium determinations of environmental samples have become quite important. Large-scale poisoning of water fowl has occurred in several watershed areas of central California. Selenium is washed or leached into these areas, helped through the widespread irrigation of selenium-containing farm soil.

Selenium is commonly found in oxidation states $(-II)$, (IV), and (VI), and as an element. $Se(VI)$ is much more stable than $Se(IV)$. The reaction chemistry of selenium is mainly that of selenide (Se^{2-}), selenite (SeO_3^{2-}), and selenate (SeO_4^{2-}). Selenious acid is a weak acid, pK_a 2.6 and 8.3. Selenious acid and selenite are much stronger oxidants than sulfurous acid and sulfite. Thus, many of the characteristic reactions are related to redox reactions, in which $Se(IV)$ can be reduced to

TABLE 8.4
SPECIATION OF TELLURIUM AND SELENIUM

Sample	Species (ret. time, min)	Eluant	Column	Detector	Detection limits	Ref.
Drinking water	Te(IV) (3), Te(VI) (12)	3 mM Na_2CO_3	Dionex Anion	suppressed conductivity	5 ppm	55
Solution	Te(IV) (6), Te(VI) (8)	6 mM Na_2CO_3	Dionex Anion	suppressed conductivity	0.5–100 ppm	56
Drinking water	Se(IV) (6), Se(VI) (18)	3 mM Na_2CO_3	Dionex Anion	suppressed conductivity	0.5 ppm	55
Waste water, copper electrode	Se(IV) (3), Se(VI) (6.5)	3.5 mM Na_2CO_3, 0.8 mM NaOH	Dionex AS4	suppressed conductivity	0.05 ppm	58
Medicines	Se(IV) (2.5), Se(VI) (5.5)	4 mM Na_2CO_3, 4 mM $NaHCO_3$	Dionex AS4	suppressed conductivity		59
Solution	Se(IV) (3.5), Se(VI) (7.6)	3 mM $NaHCO_3$, 2.4 mM Na_2CO_3	Dionex AS4	DCPAE	1 ppm	42
Soil	Se(IV) (8.2), Se(VI) (13.3)	4mM PHBA, pH 8.0	Wescan Anion/R	nonsuppressed conductivity	1 ppm	60
Kraft black liquor	Se(IV) (8), Se(VI) (17)	1 mM Na_2CO_3, 5 mM NaOH 0.8 mM p-cyanophenol	Dionex AS5	suppressed conductivity	0.01 ppm	10

elemental selenium. The acid strength of selenic acid (H_2SeO_4) is similar to sulfuric acid.

Selenium can exist in at least two different ionic forms in environmental samples: selenite (SeO_3^{2-}) and selenate (SeO_4^{2-}). The concentration and speciation of selenium in a given sample depend on the pH and redox conditions, the solubility of its salts, the biological interactions, and the reaction kinetics. For example, in sea water, SeO_3^{2-} is the dominant species, but in river or tap water selenium can be found in roughly equal Se(IV) and Se(VI) amounts [57].

Zolotov et al. [55] developed a suppressed IC method for the simultaneous determination of SeO_3^{2-} and SeO_4^{2-} in the presence of F^-, Cl^-, NO_3^-, PO_4^{2-}, and SO_4^{2-}. The separation took about 30 minutes. The method was applied to river and tap water. Sensitivity and precision were good. Hydrogen peroxide was used in the sample pretreatment to decompose organoselenium compounds so that total selenium could be determined.

The determination of SeO_3^{2-} and SeO_4^{2-} in copper electrolyte and wastewater was investigated by Noguchi et al. [58]. Some interference by Cl^- and F^- was reported. DCPAE detection was used by Urasa and Ferede [42]. Results were a 1000 times more sensitive than conductivity detection. One of the advantages of atomic emission detection was described in this work. Identical molar sensitivity was obtained for both species. Recently, Mehra et al. [60] developed a novel single column IC method to determine selenium species in seleniferous soil samples. The chromatography took about 14 minutes and there were no reported interferences. A method for the determination of selenium species after oxygen combustion has been developed and applied to bulk prednisolone [59]. However, it was found that large amounts of SO_4^{2-} can interfere.

8.3.6. Vanadium

Vanadium has a relative abundance of about 0.02%. Oxidation states of (V) to (−I) are known. Vanadium solutions generally contain several species in a complicated series of equilibria. V(V) and V(IV) are both stable, with the former mildly oxidizing and represented mainly by oxy-species. Pervanadyl ion (VO_2^+) is a major species in strongly acidic solutions, while in strong base, the mononuclear vanadate, VO_4^{3-} exists. V(IV) ions are stable in acid to give blue solutions of vanadyl ion (VO^{2+}). A number of anions of (IV) oxidation state are known including VO_3^{2-}, and $V_4O_9^{2-}$. These ions are stable under alkaline conditions.

Much of the analytical chemistry of vanadium is concerned with its use in ferrous and nonferrous metallurgy. Vanadium also finds application in catalysis and in the paint and ceramic industries. Environmental concerns about vanadium arise primarily from air pollution problems. Vanadium can be released from fly ash and oil combustion products. There are only a few references on vanadium speciation. One reference reported the simultaneous determination of V(IV) and V(V) [61]. Post-column reaction with PAR resulted in detection limits of about 10 ppb, even in the presence of high concentration of phosphate. Unfortunately, the studies were not carried out in samples. Urasa et al. [5] used DC-PAE detection to speciate VO^{2+} another vanadium species thought to be $VOCl_4^{2-}$.

8.3.7. Tin

Tin forms two stable inorganic species of Sn(II) and Sn(IV). Sn(II) is added to tin/lead alloy plating baths. The Sn(II)/Sn(IV) ratio is important to the plating bath performance. An IC separation was carried out using 0.3 mM HCl eluant [62]. Neither Sn(II) or Sn(IV) were strongly retained by the cation-exchange column used in this work. Inorganic tin speciation is quite difficult because Sn(II) will hydrolyze easily at neutral and alkaline pH.

Tin has a strong tendency to form organometallic complexes. Organotin compounds are used in marine antifoulant agents. Of course there can be many tin species when combined with organics. MacCrehan [63] reported the separation of n-butyl$_3$Sn$^+$, ethyl$_3$Sn$^+$, and methyl$_3$Sn$^+$ using cation chromatography and differential pulse amperometric detection. Organotin compounds tend to foul or poison the working electrode surface. The differential pulse technique used here eliminated this problem through the reoxidation of the reduction products on the electrode surface.

The speciation of tin in natural waters was performed by Ebdon et al. [64] using AA detection. Sn(II), Sn(IV), and butyl$_3$Sn$^+$ were separated. An extraction/preconcentration sample preparation procedure was used. Jewett and Brinckman [65] used graphite furnace AA (GFAA) to detect several diorganotin and triorganotin species. Cation exchange and reverse phase chromatography was used in the several samples listed. Chromatographic separation of both dialkyltin and trialkyltin species appear to follow cation-exchange separation mechanisms.

8.3.8. Mercury

The harmful effects of mercury and organomercury species are well known. Disastrous effects, both on personal levels and large population levels, have resulted from exposure to certain mercury species. A method for Hg(I) and Hg(II) was reported which used on-column derivatization with diethyldithiocarbamate complexing agent. This process of using a complexing reagent seems appropriate for this speciation determination. In this way the mixture can be "locked in" before the chromatography takes place. This is particularly important for metals that change oxidation states easily. A reverse phase column and UV detection at 350 nm was employed [67]. Mercury compounds all appear to absorb UV light. 254 nm detection has also been used with reasonable sensitivity and selectivity.

ICP mass spectrometry was used by Bushee [68] to speciate Hg(II) and thimerosol, a Hg containing preservative. Other work showed the separation of methylHg$^+$, ethylHg$^+$, and Hg(II). The detection limits were extremely good at about the 1 ppb level.

The addition of 2-mercaptoethanol (ME) to an eluant complexes with mercury compound to produce charge neutral compounds that can be separated on reverse phase columns. MacCrehan et al. described this procedure in a separation of Hg^{2+}, methylHg$^+$, ethylHg$^+$, and phenylHg$^+$ [70, 71].

8.3.9. Other metals

Aluminium species are oxidation state (III). In aqueous solution, the simple ion exists as Al(H$_2$O)$_6^{3+}$. This ion readily dissociates to give other ions such as Al(H$_2$O)$_5$OH^{2+}, all of which are colorless. Over a wide pH range under physiological conditions in alkaline solution, the species appear to be Al(OH)$^{2+}$, Al(OH)$_3$, Al(OH)$_4^-$, Al$_3$(OH)$_{11}^{2-}$, Al$_6$(OH)$_{15}^{3+}$, and Al$_8$(OH)$_{22}^{2+}$. Study of the substitutions of aluminum aqua ion by ligands such as SO$_4^{2-}$, citrate, and EDTA has been established by ^{27}Al NMR spectroscopy [72].

A novel study on the speciation of aluminum in solution has been reported recently by Bertsch et al. [73]. Fluoro, oxalato, and citrato aluminum complexes were identified as distinct peaks together with free Al(III). Post-column reaction/UV detection was used. These studies were used in kinetic, ion exchange, and toxicological investigations.

Gold cyanide complexes are important in gold plating baths. As the Au(III) bath content increases, the plating efficiency is decreased. Mo-

TABLE 8.5

SPECIATION OF MISCELLANEOUS METALS

Sample	Species (ret. time, min)	Eluant	Column	Detector	Detection limits	Ref.
Solution	V(IV) V(V)	8.6 mM LiOH	Dionex CS5	UV	5 ppm	61
Solution	VO^{2+} (3.5), $VOCl_4^{2-}$ or VO_3^- (1.2)	7.5 mM trilithium citrate, 10 mM oxalic acid	Dionex CS5	DCPAE	0.05 ppm	5
Solution	Sn(II) (2.5), Sn(IV) (4)	0.3 mM HCl	Dionex CS5	UV/VIS, 520 nm	1 ppm	62
Solution	n-butyl$_3$Sn$^+$ (5.5), ethyl$_3$Sn$^+$ (9), methyl$_3$Sn$^+$ (19)	42 mM NH$_4$ acetate, pH 5.3, 60/40 methanol/water	Whatman PXS Catex	amperometric	1 ppb	63
Natural water	Sn(II) (1.8), Sn(IV) (2.2), n-butyl$_3$Sn$^+$ (6)	100 mM NH$_4$ acetate 70/30 methanol/water	Partisil 10 SCX	AA	200 ng	64
Saline solution	phenyl$_3$Sn$^+$, (4.8) n-butyl$_3$Sn$^+$, (14.3) propyl$_3$Sn$^+$ (22)	methanol	Lichrosorb C8	GFAA	100 ng	65
Antifoulant leachates	various di and tri organotin	60 mM NH$_4$ acetate 30/70 methanol/water	Partisil 10 SCX	GFAA	5–30 ng	66
Solution	Hg(I) (5.9), Hg(II) (7.2)	0.05% sodium diethyl-dithiocarbonate, 70% methanol, 10% chloroform	Hypersil ODS	UV, 350 nm	0.1 ppm	67
Contact lens solution	methylHg$^+$ (6), HgCl$_2$ (8), ethylHg$^+$ (13), thimerosol (17)	60 mM NH$_4$ acetate, 0.005% 2-mercaptoethanol, 3% acetonitrile, pH 5.3	Waters C18	ICP mass spect.	1 ppb	68

Sample	Analytes	Mobile phase	Column	Detection	Detection limit	Ref.
Solution	Hg(II) (11), methylHg$^+$ (13), ethylHg$^+$ (24)	30/70 ethanol/50 mM NaBr, pH 3	Hypersil C18	UV and ICPAE		69
Solution	Hg(II) (3.9), methylHg$^+$ (4.5), ethylHg$^+$ (6.5), phenylHg$^+$ (12)	60 mM NH$_4$ acetate, 0.0005 M 2-mercaptoethanol, 40/60 methanol/water, pH 5.5	Spherosorb ODS	amperometric		70, 71
Solution	difluorotetra-aqua Al (1.7), fluoropenta-aqua Al (2.2), hexa-aqua Al (9)	700 mM NH$_4$Cl, pH 3.2	Dionex CS3	UV, 310 nm after PCR rxn with Tiron	10 ppb	73
Plating bath cyano complexes	Au(I) (5), Au(III) (11)	2 mM tetrabutylammonium hydroxide, 0.2 mM carbonate, 10–40% acetonitrile	Dionex MPIC-NSI	suppressed conductivity	0.5 ppm 77	74, 75,
Plating bath cyano complexes	Au(I) (10), Au(III) 930)	Waters PIC A, acetonitrile	Waters Rad-Pak C18	UV, 214 nm	0.1 ppm	76
Solution	methyl$_3$Pb$^+$ (3.5), ethyl$_3$Pb$^+$ (2.7), phenyl$_3$Pb$^+$	60 mM NH$_4$ acetate, pH 5.5, 40/60 methanol/water	Lichrosorb-NH$_2$	amperometric and UV	0.1 ng	71, 79, 80
Solution	Cu(I) (2.9), Cu(II) (7.8)	5 mM pentane sulfonic acid, pH 2.85	Alltech RP-18	RI or ICPAE	1 ppm	52
Solution	Mn(II) (4.6), Mn(III) (26)	1 mM oxine borate buffer, pH 9.0, 60% acetonitrile	Knauer RP-8	UV, 254 nm	1 ppm	78

bile phase ion chromatography can be used to determine total gold as well as $Au(CN)_2^-$ and $Au(CN)_4^-$ [74–76]. An anion exchange method was also reported [77]. Conductivity or UV detection can be used.

Lead speciation of $ethyl_3Pb^+$ and $methyl_3Pb^+$ along with some other organometallic species was reported by MacCrehan et al. [71]. Lichrosorb NH_2 column and an ion-pairing type eluant was used. Bushee, Krull, and coworkers described an ion-pairing reverse phase method for the separation of Cu(I) and Cu(II) [52]. Detection was by ICPAE. The effect on the retention of Cu(I) of changing the ion pairing reagent from pentane sulfonic acid to octane sulfonic acid was shown. Mn(II) and Mn(III) along with some other metals were separated on a C18 column and detected by UV [78].

An elegant approach to determine Pt, Pd, and Au as their chloro complexes using IC was reported by Rocklin [82]. UV detection was used resulting in detection limits of 0.03–0.1 ppm. In other work, IC was employed for the separation of chloro complexes of Pt, Pd, and Ir. UV detection was used with a 1 mM sulfosalicyclic acid, pH 4.2 eluant [83].

EDTA anionic complexes of Cd, Ni, Cu, and Zn has been separated by IC [84]. The results agreed well with those obtained by AA and ICP methods.

The separation and quantification of cyano complexes of various metals have been successfully carried out by Hilton and Haddad [85]. Cyano complexes of Cu(I), Ag(I), Fe(II), Fe(III), Co(II), Au(I), Au(III), Pd(II), and Pt(II) were analyzed using ion-pairing chromatography. The methods can be extended to many speciation problems.

8.4. CONCLUSION

There is much work on organospeciation that has not been reported here. For example, several metals readily form a variety of species with carbonyls. Cappon [4] wrote an excellent review of the HPLC speciation of organometallic and inorganic species. The speciation of inorganic and organometallic Sn, Pb, and Hg were described in detail. The reader is urged to consult this reference.

The reader is cautioned, that with only a few exceptions, metal speciation in one chromatographic run has been emphasized in this chapter. If there is an interest in a particular species, or if two analytical runs are acceptable, then there may be other work of interest that has been published.

A problem with metal speciation is the research itself is incomplete and has not been published. Metals can form a variety of forms, but as yet there are no good analytical procedures for separating, detecting, and quantifying many of these forms. Frequently, the act of separation is enough to change the sample. Detection methods have advanced rapidly, as shown by several chapters in this book. Perhaps using these new detection tools, chromatography and sample preparation methods can too advance.

ACKNOWLEDGEMENT

The authors are grateful to Peter Jackson for the use of his database in this work.

REFERENCES

1   M. Bernhard, F.E. Brinckman and P.J. Sadler, The Importance of Chemical Speciation in Environmental Processes, Springer Verlag, New York, 1986.
2   C.J. Cappon, LC-GC, 5 (1987), 400.
3   G.E. Batley, Trace Element Speciation: "Analytical Methods and Problems", CRC Press, Boca Raton, FL., 1989.
4   C.J. Cappon, LC-GC, 6, (1988), 584.
5   I.T. Urasa, S.H. Ram and V.D. Lewis, in Advances in Ion Chromatography, Vol 2, P. Jandik and R.M. Cassidy eds, Century International, Inc, Franklin, MA, 1990, p 93.
6   D.T. Gjerde, J. of Chromtogr., 439 (1988) 49.
7   D.T. Gjerde and J.S. Fritz, Ion chromatography, Huethig, Heidelberg and New York, 1987.
8   O. Shpigun and Yu.A. Zolotov, Ion Chromatography in Water Analysis, Wiley, New York, 1988.
9   F. Nakata, S. Hara, H. Matsuo, T. Kumamaru and S. Matsushita Anal. Sci., 1 (1985) 157.
10 Dionex Application Note TN24, May, 1987.
11 A.F. Geddes and J.G. Tarter, Anal. Lett. 21 (1988) 857.
12 H. deBeer and P.P. Coetzee, S. Afr. J. Chem. 41 (1988) 152.
13 I.T. Urasa and S.H. Nam, J. Chromatogr. Sci., 127 (1989) 30.
14 I.S. Krull, D. Bushee, R.N. Savage, R.G. Schleicher and S.B. Smith, Anal. Lett., 15 (1982) 267.
15 H. Saitoh and K. Oikawa, J. Chromatogr., 329 (1985) 247.
16 C.O. Moses, A.L. Herlihy, J.S. Herman and A.L. Mills, Talanta, 35 (1988) 15.
17 J. Weiss, Handbook of Ion Chromatography, Dionex Corporation, Sunnyvale, CA 1986.
18 V.K. Jones and J.G. Tarter, Analyst, 113 (1988) 183.
19 J.M. Wang, F.C. Chang and Y.C. Yeh, J. Chinese Chem. Soc. 30 (1983) 167.

236

20 Dionex Technical Note 23, 1987, Application Update 101, 1985, Technical Note 10R, 1985.

21 G. Schwedt, Chromatographia, 12 (1979) 613.

22 K. Mori and D. Kuroda, Kaguku to Kogyo, 59 (1985) 218, Japanese.

23 P.J. Hobbs, P. Jones and L. Ebdon, Anal. Proc. 20 (1983) 613.

24 J. Riviello, A. Fitchett and E. Johnson, Proc. Int. Water Conf., Eng. Soc. West PA, 43rd (1982) 458.

25 D. Yan and G. Schwedt, Fres. Z. Anal. Chem. 320 (1985) 325, German.

26 P.R. Haddad, P.W. Alexander and M. Trojanowicz, J. Chromatogr., 324 (1985) 319.

27 M. Takayanagi and T. Yashiro, J. Chromatogr. 374 (1986) 378.

28 J.N. Barisci, G.G. Wallace and J.M. Riviello, Chromatographia, 25 (1988) 162.

29 E.A. Boyle, B. Handy and A. Van Geen, Anal. Chem., 59 (1987) 1499.

30 D. Yan, J. Zhang and G. Schwedt, Fres. Z. Anal. Chem., 331 (1988) 601, German.

31 S.S. Heberling, J.M. Riviello, M. Schifen and A.W. Ip, Res. Dev., 74 (Sept. 1987).

32 D.F. Hilton and P.R. Haddad, J. Chromatogr., 361 (1986) 141.

33 Waters Chrom. Div. IC Lab. Report No. 272.

34 B.K. Lavine, S. McMillan, A.J.I. Ward and O. Donoghue in Advances in Ion Chromatography, Vol 1, P. Jandik and R.M. Cassidy, eds., Century International, Inc. Franklin, MA, 1989, p 195.

35 R.R. Turner, Environ. Sci. Techn., 15 (1981) 1062.

36 C.J. Craig, Organometallic Compounds in the Environment, Longman, London, 1986, p 198.

37 T. Takamatsu, M. Kawashima and M. Koyama, Bunseki Kagaku, 28 (1979) 596.

38 G.R. Ricci, L.W. Shepard, G. Coloves and N.E. Hester, Anal. Chem. 53 (1981) 610.

39 R.J. Williams, Anal. Chem. 55 (1983) 851.

40 T. Jupille in Ion Chromatography, J.G. Tarter ed., Marcel Dekker, New York, 1987, p 79.

41 L.K. Tan and J.E. Dutrizac, J. Chromatogr., 405 (1987) 247.

42 I.T. Urasa and F. Ferede, Anal. Chem., 59 (1987) 1563.

43 Z. Hu and Y. Tang, Analyst, 113 (1988) 179.

44 E.C.V. Butler, J. Chromatogr., 450 (1988) 353.

45 E. Papp and A. Fehervari,. J. Chromatogr., 447 (1988) 315.

46 N. Hirayima and T. Kuwamoto, J. Chromatogr., 457 (1988) 415.

47 G.K.-C. Low, G.E. Batley and S.J. Buchanan, Anal. Chim Acta, 197 (1987) 327.

48 G.K.-C. Low, G.E. Batley and S.J. Buchanan, Chromatographia, 22 (1986) 292.

49 K.E. LaFreniere, V.A. Fassel and D.E. Eckels, Anal. Chem., 59 (1987) 879.

50 B.S. Chana and N.J. Smith, Anal. Chim, Acta, 197 (1987) 177.

51 L.K. Tan and J.E. Durtizac, Anal. Chem., 58 (1986) 1383.

52 D. Bushee, I.S. Krull, R.N. Savage and S.B. Smith, Jr., J. Liq. Chromatogr., 5 (1982) 463.

53 R.A. Stockton and K.J. Irgolic, Inter. J. Environ. Anal. Chem., 6 (1979) 313.

54 F.E. Brinckman, K.L. Jewett, W.P. Iverson, K.J. Irgolic, K.C. Ehrhardt and R.A. Stockton, J. Chromatogr., 191 (1980) 31.

55 Yu.A. Zolotov, O.A. Shpigun, L.A. Bubchikova and E.A. Sedd'nikova, Dokl, Akad. Nauk. SSR, 263 (1982) 889, Russian.

56 S.G. Chen, K.L. Cheng and C.R. Vogt, Mikrochim Acta, (1983) 473.

57 R.J. Shamberger, Biochemistry of Selenium, Plenum, New York, 1873, p 185.

58 F. Noguchi, H. Ito, T. Nakamura, Y. Ueda and K. Ota, Anal. Sci., 4 (1988) 398.

59 M. Murayama, M. Suziki and S. Takitani, Anal. Sci., 4 (1988) 585.

60 H.C. Mehra and W.T. Frankenberger, Jr., Chromatographia, 25 (1988) 585.
61 R.E. Smith, Ion Chromatography Applications, CRC Press, Boca Raton, FL 1988.
62 R.A. Cochrane in Trace Metal Removal from Aqueous Solution, R. Thompson, Ed., The Royal Society of Chemistry, London, 1986, p 197.
63 W.A. MacCrehan, Anal. Chem., 53 (1981) 74.
64 L. Ebdon, S.J. Hill and P. Jones, Analyst, 110 (1985) 515.
65 F.E. Brinckman, W.R. Blair, K.S. Jewett and W.P. Iverson, J. Chrom. Sci., 15 (1977) 493.
66 K.L. Jewett and F.E. Brinckman, J. Chrom. Sci., (1982).
67 R.M. Smith, A.M. Butt and A. Thakur, Analyst, 110 (1985) 35.
68 D.S. Bushee, Analyst, 113 (1988) 1167.
69 C.H. Gast, J.C. Kraak, H. Poppe, F.J.M.J. Maessen, J. Chromatogr. 185 (1979) 594.
70 W.A. MacCrehan, R.A. Durst, Anal. Chem., 50 (1979) 2108.
71 W.A. MacCrehan, R.A. Durst and J.M. Bellama, Anal. Lett., 10 (1977) 1175.
72 L.O. Ohman and S. Sjonberg, J. Chem. Soc. Dalton Trans., (1983) 2513.
73 P.M. Bertsch and M.A. Anderson, Anal. Chem. 61 (1989) 535.
74 Dionex Application Note 40R (1983).
75 J. Weiss, CLB, Chem. Labor Betr 35 (1984) 59.
76 Waters Chrom. Div. ILC Series Application Brief No. 5001.
77 M. Nonomura, Met. Fin. 85 (1987) 15.
78 B.W. Hoffman and G. Schwedt, J. HRC and CC, 5 (1982) 439.
79 M. Blaszkiewicz, G. Baumhoer and B. Neidhart, Z. Anal. Chem. 317 (1984) 221.
80 D.K. Orren, J.C. Caldwell-Keukel and P. Mushak, J. Anal. Toxicol. 9 (1985) 258.
81 W.T. Frankenberger, Jr., H.C. Mehra, and D.T. Gjerde, J. Chromtogr. 504 (1990) 211.
82 R.D. rocklin, Anal. Chem. 56 (1984) 1959.
83 O.A. Shpigun and Yu. E. Pazuktina, Zh. Analit. Khim. 42 (1987) 1285.
84 T. Takana, Z. Anal. Chem. 320 (1985) 125.
85 P.R. Haddad and N.E. Rochester, J. Chromatogr. 439 (1988) 23.

I.S. Krull (Ed.), *Trace Metal Analysis and Speciation*
Journal of Chromatography Library Series, Vol. 47
© 1991 Elsevier Science Publishers B.V., Amsterdam

239

Chapter 9

Interfacing of GC/HPLC with Direct Current Plasma (DCP) Emission Spectroscopic Detection for Trace Metal Analysis and Speciation

IRA S. KRULL [1,] * and Wm. CHILDRESS [2]
[1] *Department of Chemistry and The Barnett Institute (341MU), Northeastern University, 360 Huntington Avenue, Boston, MA 02115, U.S.A.*
[2] *Analytical Chemistry Branch, U.S. Food and Drug Administration, Winchester Engineering and Analytical Chemistry, 109 Holton Street, Winchester, MA 01890, U.S.A.*

9.1. INTRODUCTION

It has become eminently clear in the past decade that various forms of atomic emission spectroscopy are well suited for chromatographic interfacing [1–9]. Several reviews and textbooks have recently appeared or will shortly appear attesting to the growing interest and applications of atomic emission spectroscopic (AES) detection in various forms of chromatography. This current text contains chapters describing the successful interfacing of gas (GC), liquid (HPLC), and supercritical fluid chromatography (SFC) with microwave induced (MIP), direct current (DCP), alternating current plasma (ACP), atomic fluorescence (AFS), and inductively coupled plasma (ICP) emission detection methods. Though MIP was perhaps the earliest and immediately successful plasma detection method utilized in GC, it was readily apparent that it might not be an ideal, stable, robust detector for HPLC, other than perhaps microbore. There are discussions elsewhere in this text regarding the current state-of-the-art in HPLC-MIP interfacing and applications. ACP appears to be a very viable alternative to the MIP as an element specific detector for both GC and HPLC [10]. The nitrogen afterglow detector introduced by V. Fassel of Iowa State University in the mid-80s, is yet another example of an emission based, element specific detector for chromatography [11]. In the 1970s, a British firm commercialized a GC-MIP system for performing trace metal speciation of volatile

---

* Author to whom correspondence and reprint requests should be addressed.

organometals [12]. This was followed in the mid-80s by the Hewlett Packard introduction of another variation of a GC-MIP system, which has gained significant attention and applications [13–15].

Commercial, hyphenated chromatography-element selective detection (ESD) systems have been few and far-between. One such system was introduced almost a decade ago by Perkin-Elmer, which utilized a carousel (merry-go-round) interface for collecting, batch-wise, fractional eluents from an open column or HPLC system and routed each of these, step-wise, to a graphite furnace atomic absorption (AAS) system [16–18]. This was an example of an off-line, indirect approach to introducing chromatographic effluents into an element specific detector, as opposed to a direct connection with continuous, automated transfer of an effluent into a flame, furnace, or plasma source. There are serious disadvantages, especially time constraints, in using any indirect, off-line approach, wherein an effluent has to be divided up into separate fractions, followed by a step-wise introduction of each fraction into the ESD. This also leads to a histogram type of data presentation, rather than the more conventional, smooth peak shapes customary in GC/HPLC today. Thus, though commercially available for some time, HPLC-AAS (atomic absorption spectroscopy), using either graphite furnace of flame methods, has not become as popular and widely employed as might be imagined [19–21]. At the present writing, it does not appear that many other commercially interfaced GC/HPLC-ESD systems have appeared on the market. This may be due to the relative triviality of the interfacings that have already been described in GC/HPLC-ICP/DCP systems. Elsewhere in this volume are described various approaches, as well as GC/HPLC-ICP/MS systems, which have all used home-made interfaces, all of which appear to have been satisfactory and very successful. In general, most flow rates for analytical GC/HPLC appear compatible with sample introduction requirements for ICP/DCP detectors. This may also be true of the atomic fluorescence emission system, but much less has been published on its interfacing than for plasma emission or atomic absorption detection [1–9].

For several years, the Winchester Engineering and Analytical Center (WEAC) of the U.S. Food and Drug Administration (FDA) has evaluated and applied a variety of GC/HPLC-DCP approaches for true element specific and selective detection. Element selective, by and large, suggests that the response is relatively unique for a certain/specific element, but that other interferents may also produce similar responses. Element specific, by and large, suggests that the response on that particular detector is *unique* for that particular element, whatever its species or form.

The WEAC/FDA facility was fortunate in the early-mid 80s in having a Spectrametrics Spectraspan IIIb model DCP instrument. Until the early 80s, it was routinely utilized for total metal determinations, and had no history of being interfaced with any chromatographic system. Together with both flame and furnace AAS, the DCP was a workhorse within FDA for performing routine or semi-routine total metal determinations in a wide variety of import/domestic type samples.

The situation of determining total metal content rapidly changed after the Japanese minimata incident, followed by the worldwide realization that methylmercury could be formed in vivo from appropriate inorganic mercury cations/species. Total Hg determinations in fish samples became less and less important, especially in court cases, when it was quickly realized by importers and their attorneys, that unless MeHg was shown to be present, total Hg content could not be used as a legal basis for seizure of imported fish species and their restriction from the U.S. markets. Thus, a wide variety of element specific analytical methods were developed, initially using AAS detection, until it was soon realized that plasma techniques offered significant advantages. The main problem, however, was the fact that in the early 80s, relatively few laboratories within FDA had plasma emission detectors/instruments, and most had only FAA/GFAA systems. This was considered, then, a serious impediment to the more widespread, general utilization of DCP as an element specific detector in GC or HPLC.

As another example of the need for trace metal speciation, as opposed to total metal content, current U.S. law allows for the addition of selenium to animal feeds and feed premixes as either selenite or selenate. Thus, any monitoring program would need to be able to determine which of these species was/were present [76]. Similar federal and/or state regulatory requirements now exist for other organometal species, such as methylmercury or tributyltin (TBT).

9.2. BACKGROUND TO INTERFACING OF CHROMATOGRAPHY WITH DCP

Nevertheless, it was clear from the prior literature, especially the work of Uden et al., that the DCP could be an almost ideal element specific detector for various forms of chromatography [22–24]. What has limited its general popularity has been its single source availability, and the fact that there has usually been a single supplier of the instrumentation and replacement parts, at least until very recently.

Today, there are other suppliers of torches, electrodes, spray chambers, and ancillary parts and supplies for most older and current DCP models. Spectrametrics, the original firm that manufactured and marketed DCP instruments, held all crucial patents and licenses, and thus had a virtual monopoly in the marketplace [25–28]. Leeman Laboratories, Lowell, MA, in the mid-80s, introduced another variation on the DCP, now using a braided graphite belt as the method of sample introduction [29, 30]. However, this instrument did not capture a large portion of the DCP market, and still today, the latest Model VI of the Spectrametrics product line holds sway and influence. There are no real competitors in the marketplace. This has, by and large, restricted a more widespread dissemination and application of DCP instrumentation for chromatographic interfacing. Indeed, most of the publications describing GC/HPLC/SFC-DCP have come from academic and/or government type laboratories [7, 22–24, 31–35]. There are far fewer literature reports of chromatography-DCP interfacing and applications than for the more readily available ICP systems. This may continue for several more years, or at least until the Spectrametrics patents no longer are in effect.

DCP appears to offer certain advantages and opportunities as a chromatography detector, in all forms of chromatography, especially when compared with ICP. It appears to be a more stable plasma for long-term HPLC interfacing, especially with the introduction of organic/aqueous eluents. The ICP appears to have a lower tolerance for organic solvents, and though carbon buildup on the DCP electrodes can become problematic with organic/aqueous solvents, there are numerous examples of normal phase HPLC-DCP interfacing with long-term usage [36–38]. ICP appears to have serious problems in tolerating high organic content reversed phase solvents in HPLC. Solvent problems probably disappear in interfacing both GC and SFC with ICP, and there are numerous references wherein both forms of AES have been successfully utilized. The actual interfacing in GC is relatively trivial. In HPLC, on the other hand, interfacing with ICP involves the conventional nebulizers used for direct ICP total metal determinations. Thus, virtually all commonly available nebulizers, including concentric tube, cross-flow, Meinhard, ultrasonic, direct injection, and electrothermal vaporization [39] have all been utilized and described for HPLC-ICP interfacing. In the area of HPLC-DCP, most publications have utilized the commercial spray chamber on the DCP itself, adding a separate transfer line from the end of the HPLC column directly into the gas flow tubing to the DCP spray chamber.

Efficiency of aerosol generation and transfer to the DCP plume region is not enhanced in the HPLC mode, which is traditionally low, and thus HPLC-DCP detection limits, in the main, are much higher than for direct-DCP determinations. This same problem arises in HPLC-ICP interfacing, where inefficient aerosol transfer, combined with chromatographic band dispersion phenomena (intra- and extracolumn), all adversely contribute to the final, interfaced detection limits. ICP/MS detection appears to overcome the problems in HPLC-AES interfacing, wherein this is now able to provide 2–3 orders of magnitude lowering of detection limits in comparison to HPLC-ICP/DCP methods, with or without hydride generation [40–45]. It does not appear that direct HPLC-ICP/DCP methods of analysis will soon provide detection limits in the trace range of current HPLC-ICP/MS approaches. Thus, if minimum detection limits (MDLs) are a crucial area of concern in any specific application of chromatographic interfacing, then one ought to immediately consider ICP/MS detection first. On the other hand, if such instrumentation, costly as ever, is not readily available, and/or if MDLs are not crucial for a given application, which is often the case, then there is every reason to hold out hope for some form of HPLC-ICP/DCP approach. The use of continuous, on-line hydride generation in HPLC-ICP/DCP, and/or sample introduction via electrothermal vaporization (commercially available), can often significantly improve final MDLs in the interfaced mode [46–49].

Other than for the problematic area of MDLs, DCP appears to be an ideal element selective detector for various forms of chromatography, and there is little question today that its application and usage will only increase. There is no need for any commercial interface, the current, commercial spray chamber and nebulizer on the DCP systems are easy enough to interface with common (reversed/normal phase) HPLC eluents, at most analytical flow rates. Ion chromatographic (IC) conditions are also quite easily interfaced, and indeed, the DCP appears to be an almost-ideal element detector for IC applications. It is tolerant of high organic containing eluents, high aqeuous eluents, high or low flow rates, buffered media, inorganic/organic salts, ion-pairing reagents, hydride generation conditions, and shows little memory effects from run-to-run. It can produce multi-element spectroscopic data for any particular chromatographic run, although most units do not come with echelle/polychromator capabilities. Thus, it is currently much easier to follow a single wavelength emission in a given chromatographic run, and then change the DCP conditions for subsequent elemental analyses. This is perhaps one of the disadvantages of current DCP systems, especially when compared with multi-wavelength (slew-scan or poly-chromator) ICP systems. However, most real world applications require determinations for a specific elemental component in a given sample, rather than for several elements and their species possibly present at the same time. The GC/HPLC-DCP systems can be operated on a routine basis for long times, with very little noise increase, instability, shutdown, or changes in the chromatographic signal-to-noise (S/N) ratio and baseline stability.

We describe here our experiences with the interfacing of both GC and HPLC to DCP, always using the Spectrametrics SpectraSpan IIIb instrument, operated in the active diagnostic mode. We discuss the various approaches to interface GC and HPLC, optimization of operational conditions, final optimized conditions, compatible HPLC mobile phases and flow rates, and then specific applications to real world, practical samples of importance to the FDA and others. Almost all of these results have appeared in print, except for the most recent work with Se speciation via HPLC-DCP methods [50]. In each area, a specific, practical application of the final technology and interfaced instrumentation was emphasized and shown practical. Overall methods, including sample preparation, clean-up, separation-detection, quantitation, and final sample studies, validation of the final results was demonstrated. Accuracy and precision were always demonstrated, as well as within-day and day-to-day reproducibility. All too often, publications presenting new detection methods will limit applications to designer chromatograms, that is, the separation and detection of prepared mixtures of stock chemicals with known retention times and responsiveness. Particularly for detectors that are touted as being extremely selective, it is important to demonstrate their capabilities with "real" samples, namely, the analytes for which they are designed in the matrices that they are typically found [51]. Rigorous quantitative comparisons are rare. Such work, and an excellent agreement between methods, adds needed credibility to claims that detection methods are suitable for routine analysis [51].

9.3. GC/DCP FOR TRACE METAL SPECIATION

9.3.1. Background and discussion

There have been several efforts to interface GC with DCP for trace metal speciation, but no commercial system has ever been introduced. Much of the earlier work in GC-DCP, especially that by Uden's group at the University of Massachusetts at Amherst, utilized a heated capillary transfer line from the exit of the GC unit to the plasma region of the DCP unit [54–58]. In this work, as in most, the effluent from the GC was routed via a rather long interface, maintained at a constant temperature, into the DCP plasma/plume region. It was not clear how much, if any, additional band broadening was actually contributed by this lengthy interface between the separation and detection instruments.

9.3.2. The determination of methylmercury (MeHg) in fish

In our own efforts in GC-DCP areas, we have taken two different approaches. In the first, that involving the determination of methylmercury in fish, the GC was interfaced very simply and inexpensively [59]. That is, a simple, isothermal, low-cost, home-made GC was constructed which could be dedicated to the DCP, allowing for routine qualitative and quantitative determinations of organomercury species in complex food matrices. This particular instrument utilized packed columns, of rather short dimensions, which could resolve methylmercury (MeHg) from the solvent front and any ethylmercury (EtHG), if present, in under 5 minutes. The interfacing itself involved a quartz jet tube to convey the GC effluent directly into the DCP plume. This particular interface allows for prolonged use without replacement. Optimization of the GC-DCP interface was accomplished, in terms of emission wavelength response, viewing height, carrier gas flow rate, voltages across the electrodes, and other experimental parameters. This was followed by a determination of detection limits, the linearity of the calibration graph, and then a comparison of actual results in GC-DCP with those obtained by GC–electron capture detection (ECD). Another comparison of the GC-DCP quantitative data for MeHg in fish samples was made via total mercury determinations using cold-vapour atomic absorption spectrometry.

Figure 9.1 indicates a schematic diagram of the SpectraSpan IIIb DCP model used in certain of these GC/HPLC-DCP studies. For the MeHg work, the dual detector arrangement consisted of ECD/DCP, while for the organotin studies (see below), the dual detector arrangement utilized an FPD/DCP combination. It was possible, as suggested by Fig. 9.4, that all three detectors (ECD/DCP/FPD) could be in place at all times, and for specific applications, output from any two of these could be monitored simultaneously. There are, as we have suggested elsewhere, serious opportunities and advantages inherent in multiple detection in GC [60]. Figure 9.2 shows the DCP's plasma/plume region,

References pp. 282–287

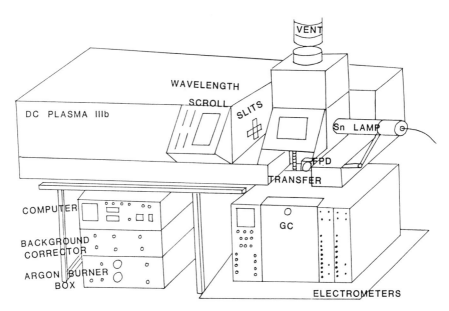

Fig. 9.1. Schematic diagram of the DCP instrument in operation. (Reprinted with permission of the copyright owner, Longman Group, UK and *Journal of Applied Organometallic Chemistry* [53].)

wherein all GC effluents were directly introduced. Figure 9.3 illustrates a schematic diagram of the GC-DCP instrumentation that was finally assembled for these particular MeHg studies. The laboratory made gas chromatograph consisted of a 4 ft × 2 mm i.d. all-glass column packed with 5% DEGS-PS on 100–120 mesh Supelcoport, operated isothermally at 155C, with an injection port temperature of 175C, and an argon flow rate through the column of 100 ml min^{-1}. The DCP quartz jet inlet tube was maintained at 190C with external heating tape. The column was first conditioned each day with one 20 μl injection of a 1000 ppm solution of mercury(II)chloride, following the AOAC protocols for analysis of MeHg in fish samples via GC-ECD [61]. The total costs for construction of this isothermal GC were about $250 or less, and it could be entirely dedicated to full-time interfacing with the DCP just for MeHg in fish samples.

A very large number of fish samples were analyzed using this instrumental methodology, with a high degree of success, reproducibility, accuracy, and overall precision. The GC sat directly underneath the DCP plasma region, without lifting the DCP off the laboratory bench surface. That is, there is sufficient dead space in the SpectraSpan IIIb unit so

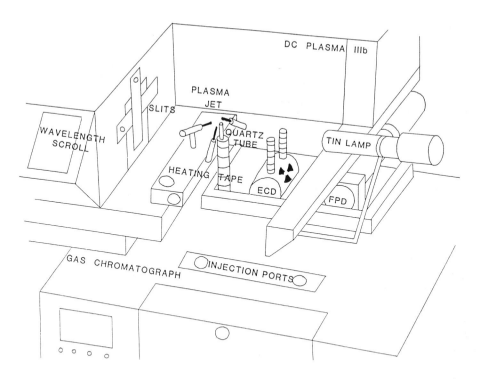

Fig. 9.2. Schematic diagram of the DCP spray chamber (nebulizer) and plasma plume region. (Reprinted with permission of the copyright owner, Longman Group, UK and *Journal of Applied Organometallic Chemistry* [53].)

that the end of the GC column and oven could sit right underneath the plume region (Fig. 9.3). The outlet of the GC column had a 1/4-in stainless steel Swagelok 90 degree elbow connection with a 1/4-in o.d. × 1 mm i.d. quartz tube for transfer of the column effluent to the DCP plume region. The top of the flat-tipped quartz tube was ground into a cone to project the tube higher into the plume. The quartz tube was 6-in long and was wrapped with heating tape to maintain the temperature at 190C, with the column maintained at 155C. The specific details for the construction of this low-cost, isothermal GC can be found in the original publication [59].

There were certain seemingly obvious advantages to this particular GC-DCP interfaced arrangement, especially the ability to avoid any heated transfer lines. That is, as soon as the chromatographic effluents left the GC column, they were immediately transfered into the plasma plume region, and thus there was virtually no lag time. There was

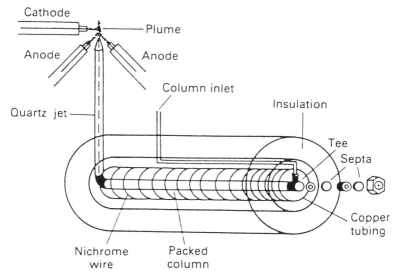

Fig. 9.3. Schematic diagram of interfacing GC with DCP for performing volatile MeHg speciation. Isothermal GC located directly beneath the DCP plasma region. (Reprinted with permission of the copyright owner, The Chemical Society (UK) and *The Analyst* [52].)

also no additional band dispersion via this interface, and even with the very short, packed columns used here, peak shapes for standard MeHg and EtHg were perfectly symmetrical with no tailing. Figure 9.4 illustrates some typical GC-DCP chromatograms of actual swordfish sample extracts using a newly developed extraction method, with an overall MeHg retention time of under 3 mins. Figure 9.4 illustrates peak shapes for standard MeHg at an injected level of about 3.0ng (3.0 μl of a 1.0 μg ml^{-1} Hg sample). Signal-to-noise levels here are extremely good. The actual swordfish samples are also shown in Fig. 9.2, now at a level of 8.0 μl injected of the final 5ml extract of the fish. Again, peak shapes for both the standard MeHg and that present in actual swordfish samples were very symmetrical and homogeneous. The detection limit for standard MeHg solutions was determined by injecting that concentration of MeHg (0.3 ng per 8 μl) which would produce a signal-to-noise ratio of about 3 : 1. Statistically, for 10 replicate injections, the analyte produced an average signal of 7.6mm with a relative standard deviation (RSD) of about 5.8%. The background noise level was 2.5mm (RSD 12.8%). A calibration graph was constructed which was linear from 0.2 to 20 μg per 8 μl injection, with a correlation coefficient of 0.9999. It is possible that linearity would extend beyond

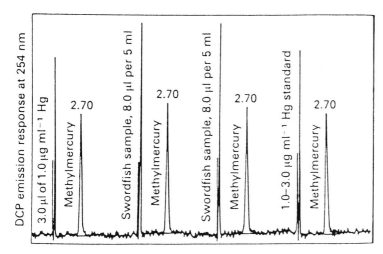

Fig. 9.4. GC-DCP chromatograms of standard MeHg and actual swordfish sample extract containing incurred levels of MeHg. No other Hg containing species were evident. (Reprinted with permission of the copyright owner, The Chemical Society (UK) and *The Analyst* [52].

this range, but for the purposes of these determinations in fish, this linear dynamic range was adequate.

Figure 9.5 illustrates a direct comparison of both GC-ECD and GC-DCP chromatograms for the same NBS-RM50 tunafish sample. It is clear that the DCP results provide improved MeHg specificity in comparison to the quite selective, but much more non-specific ECD detection capabilities, at least for these particular samples. A large number of extraneous peaks are evident in the GC-ECD chromatogram, virtually all of which have disappeared in the GC-DCP one. Solvent fronts in GC-ECD are also considerably broader and more pronounced than those in GC-DCP, thereby increasing (higher) the detection limits for MeHg possible using certain extraction solvents. The overall simplicity of the GC-DCP chromatogram for an actual fish extract (Fig. 9.5), in comparison with GC-ECD, is striking, leading to improved analyte identification and an easier quantitation with no interferences.

The overall GC-DCP results for MeHg in fish samples demonstrated accurate, precise, and reproducible determinations, day-to-day and within-day. A comparison of total Hg levels in the same fish samples has shown that, in general, they are higher than those determined by GC-DCP, which only showed one Hg containing species, MeHg. There may very well be other, perhaps nonvolatile Hg species, such as inorganic Hg, which would not gas chromatograph. Additional sample preparation

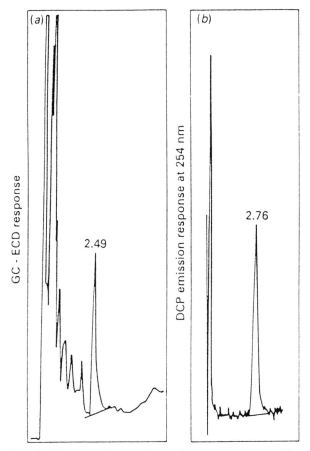

Fig. 9.5. A comparison of two chromatograms for the same MeHg in NBS-RM50 tunafish sample. Left chromatogram, that for GC-ECD detection, right chromatogram, that for GC-DCP detection at 254 nm. GC conditions were slightly different for GC-ECD vs GC-DCP, leading to slight differences in retention times (min). (Reprinted with permission of the copyright owner, The Chemical Society (UK) and *The Analyst* [52].)

was necessary for GC-DCP, especially the need to pre-concentrate the fish extracts before injection. This was because the final detection limits by GC-DCP are generally poorer (higher) than via GC-ECD, which thus does not require such pre-concentration steps. However, there was (Fig. 9.5), a serious trade-off in analyte specificity. The DCP, being Hg-specific, avoids almost all possible interferences, and thus can provide more accurate determinations of MeHg. The advantages of element selective or specific detection for MeHg or other Hg containing species, especially by GC, are therefore obvious.

9.3.3. The determination of organotins in fish and shellfish

Just as MeHg has become a widely distributed and commonly found environmental pollutant in many fish species, especially swordfish and tunafish, so too has tributyl tin (TBT) gained in recognition as an even more widely distributed aquatic pollutant and toxicant [53]. TBT can now be found in almost all fish/shellfish samples taken anywhere in the world, even from native salmon in Scotland, Denmark, Norway, or elsewhere. The U.S. Food and Drug Administration (FDA) has, in part, a responsibility to continuously monitor foods imported into the U.S. destined for human consumption. It was, in part, the purpose of this study to develop improved GC-element selective detection (ESD) approaches for organotins in fish and shellfish. And, to then utilize such newer, validated approaches for as many real world marine samples as possible and practical.

By and large, most currently employed approaches for the accurate and precise quantitation of TBT and other butyltins rely on GC with some type of element "selective" or specific detection technique. Most of these use flame photometric detection (FPD) with a tin specific filter at 600nm emission. Though somewhat selective for tin containing species, it is not 100% specific for tin alone. Thus, the combination of GC with FPD and DCP appeared to be a very reliable and practical approach to obtain one and/or two "selective" chromatograms from one or two injections of a fish or shellfish extract.

Most literature reports have utilized some type of pre-injection derivatization for TBT and the other butyltin species. Such approaches have generally used hydridization or alkylation, in order to provide improved GC performance characteristics for the original species. Improvements in the off-column derivatization methods generally reported have quite recently used reaction GC to form the hydrides of TBT and its analogs prior to FPD detection [62–64]. This remains an entirely feasible, continuous, and automatable approach to perform cn-column reactions in GC, especially for compounds such as TBT which can readily form metal/element hydrides.

There have been very few reports on the direct GC determination of butyltins without some type of pre-column derivatization [65]. In general, those reports that used direct injection, have shown broadened peak shapes, serious tailing of peaks, poor column efficiencies, and less-than-ideal peak resolution or capacity. An ideal GC-element selective detection (ESD) method might involve the following steps: (1) simple sample work-up and extraction from fish/shellfish with high recovery efficiencies; (2) no artifact formation of the butyltin species of interest during sample work-up or GC-ESD; (3) no prior derivatization off-line or on-line before GC injection; (4) good chromatographic performance properties with modern fused silica capillary or megabore columns; (5) high analyte selectivity and identification via dual element selective detectors (FPD/DCP); (6) element ratios obtainable via multi-wavelength (element) identification; (7) good accuracy, precision, and reproducibility of quantitations; and (8) high sample throughput and fast turn-around times.

We have reported an analytical method for volatile organotins that utilized direct sample extract injection onto a fused silica, megabore GC column containing a thin film (1 μm) of the immobilized stationary phase on the column walls. Separation was followed by either FPD and/or DCP detection, in order to provide one/two chromatograms whose peak heights and/or areas could be used for subtraction or ratioing to further improve analyte identification. It is also feasible to make a single injection and obtain simultaneously both FPD/DCP chromatograms. The overall method, GC-FPD/DCP, was optimized with regard to standard analytical figures of merit. Validation was derived from single blind spiked fish samples, as well as recovery studies on several real world samples. The FPD and DCP results have, in general, been in excellent agreement. Finally, the overall methodology was applied to as many fish and shellfish samples as obtainable, food destined for the marketplace and eventual human consumption.

We have not found any literature reports that have placed the DCP plume directly above the GC column exit, other than that described above [52]. In our earlier study, emphasizing MeHg, we had constructed a low-cost, independent, isothermal GC that slipped directly under the DCP (Fig. 9.3). In the current studies, we actually lifted the DCP over the GC column exit (Figs. 9.1 and 9.2), so that the effluent directly entered the DCP plume region. This final instrumental arrangement, will allow us to split the GC effluent, and to obtain, if desired, both FPD and DCP chromatograms simultaneously from a single injection. A variable ratio splitter would/could provide different fractions directed to each detector and thus different ratios of peak heights/areas. This could provide additional analyte confirmation using different detector peak height/area ratioing on a single injection.

Figure 9.2, with the DCP top removed, illustrates the arrangement of the column exit, DCP plume region, and location of the two other detectors available with this GC system, ECD and FPD. For these studies, we used but a single detector per injection, and all the chromatograms depict a single detector's response, FPD or DCP. Future studies will emphasize a variable ratio, post-column effluent splitter, providing simultaneous dual chromatograms, FPD/DCP (organotins) or ECD/DCP (organomercury determinations). Placing the DCP above the GC column exit has provided for virtually zero extracolumn variance, there is but a short segment of narrow i.d., heated quartz tube to transfer the column effluent directly into the DCP plume.

All four organotins of current environmental or FDA interest have been baseline resolved, without prior derivatization, using the GC-FPD conditions indicated in Fig. 9.6 [53]. Similar chromatograms, not

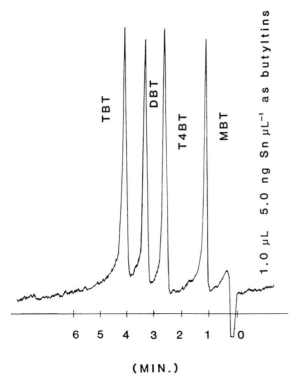

(MIN.)

Fig. 9.6. GC-FPD chromatogram for a mixture of four standard organotins under optimized GC conditions: megabore 1 μm DB-17 column, 30 m × 0.53 mm, operated at 160°C, FPD at 220°C, injection port at 200°C, flow rate 5 ml min^{-1} helium. Melpar FPD unit: tin filter at 600 nm. FPD gas flows: hydrogen 200 ml min^{-1}; air 120 ml/min; oxygen 20 ml/min. 1.0 μl 0.1 mol m^{-3} HCl in acetone followed injection, in same syringe. *TBT* = tributyltin; *DBT* = dibutyltin; *T4BT* = tetrabutyltin; *MBT* = monobutyltin. (Reprinted with permission of the copyright owner, Longman Group, UK and *Journal of Applied Organometallic Chemistry* [53].)

References pp. 282–287

254

depicted here, have been obtained via GC-DCP. A 1 μm film thickness of DB-17 in a megabore column of 0.53 mm i.d., 30 m long, was able to produce the separations and peak shapes indicated. We suspect that this has been possible because of the thin film thickness now available with immobilized, wall coated open tubular capillary (megabore) type columns. Though some peak asymmetry is evident, perhaps due to the FPD itself, all four peaks are baseline resolved within 5 mins. It has proven crucial that the standard organotins and sample extracts were made up in 0.2N HCl in acetone, in order to improve peak shape and reproducibility of injections. In the absence of the added HCl, both peak shape and resolution suffered.

Because TBT and DBT are the two most commonly found organotins in the environment, we have emphasized their analytical figures of merit. Figure 9.7 illustrates a typical GC-DCP chromatogram for single

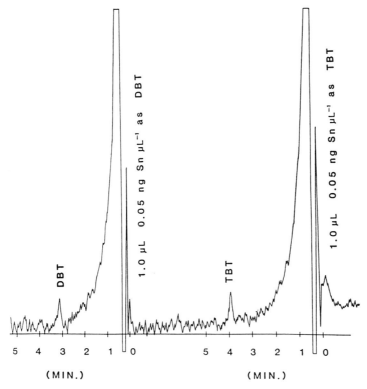

Fig. 9.7. GC-DCP chromatogram of tributyltin and dibutyltin at the limits of detection, with conditions as in Fig. 9.6. (Reprinted with permission of the copyright owner, Longman Group, UK and *Journal of Applied Organometallic Chemistry* [53].)

injections of 0.05 ng Sn as DBT and 0.05 ng Sn as TBT, 1 μl injections of each on-column. Though baseline noise is considerable, each peak is obvious and peak shape is adequate for quantitation at these levels. DBT or TBT could both be detected with a low %RSD at the 0.05 ng level of Sn. Such levels represent a concentration injected of 50 ppb (parts per billion). TBT can easily be detected with a low %RSD at the 0.25 ng level of Sn as TBT using GC-DCP, with a signal/noise ratio of 3:1. This level again corresponds to a concentration of 0.05 ng μl^{-1} of Sn as TBT, or 50 ppb.

Calibration plots were obtained for both DBT and TBT via GC-FPD and TBT via GC-DCP, ranging from detection limits, as above, up to and including 5 or 100 ppm concentrations. The data points fit the equation for a straight line in each case, with correlation coefficients and coefficients of determination as presented elsewhere [53]. Though not demanding a straight line for such plots, the data is entirely consistent with linearity, suggestive of good quantitations possible over the concentration ranges studied.

Figure 9.8 illustrates a typical GC-DCP chromatogram for TBT with conditions indicated. Except for a different carrier gas and increased flow rate, the GC conditions were identical to those for GC-FPD. The retention time for TBT was about half that in the GC-FPD chromatograms, mainly because of the higher carrier gas flow rate. Under these particular GC conditions, TBT was not adequately resolved from other possible tin containing species. Since we first proved that only TBT was present via the GC-FPD results, there was no need to use GC-DCP conditions that would also separate TBT from tin species, not present in any of these particular samples. However, if needed, we have shown that the other three normally occurring organotins (DBT, MBT, T$_4$BT) all separate from TBT under GC-DCP conditions approaching those used in GC-FPD. The final peak shape and asymmetry factor for TBT via GC-DCP were somewhat improved over those found by GC-FPD analysis (Fig. 9.6). This may have been due to the differences in detector response times, dead volumes, or contact time for FPD vs DCP. It may also have to do with the faster flow rate and shorter retention time in the GC-DCP mode.

In order to validate, in part, this newer approach, both sample work-up and GC-FPD/DCP detection, we have performed several separate, single blind spiking studies with different levels of TBT. Table 9.1 summarizes some of this data, wherein two flounder (fish) and one whiting (fish) sample were separately spiked by one analyst, at the levels indicated, and these were then analyzed by GC-FPD a second

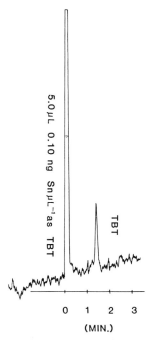

Fig. 9.8. GC-DCP chromatogram for a single injection of standard tributytin, at the level indicated, under optimized GC-DCP conditions, megabore 1 μm DB-17 column, 30 m × 0.53 mm, operated at 150°C, injection port 180°C, column flow rate 35 ml min^{-1} argon, DCP at 303.41 nm, gain setting 40, PMT 9, sleeve pressure 50 psi (345 kla) active diagnostic mode, recorder chart speed 0.5 in min^{-1} (1.3 cm min^{-1}, 10 mV FSD). (Reprinted with permission of the copyright owner, Longman Group, UK and *Journal of Applied Organometallic Chemistry* [53].)

analyst, levels unknown. Percent recoveries represent the agreement between the spiked and found levels, which ranged from 92.6 to 98.6%. Reproducibility of these determinations were quite good, with standard deviations of ±1.08–3.60 for levels spiked below 200 ppb in TBT. These particular samples of fish contained no incurred levels of TBT [53]. Table 9.2 compares GC-FPD and GC-DCP quantitative data for three separate fish samples, now containing incurred levels of TBT, as indicated. It is clear from the data in Tables 9.1 and 9.2 that in all cases, GC-DCP and GC-FPD results agreed very well, indeed they were always within experimental error. In addition, the single blind spiking results via GC-FPD were in agreement with actual levels of TBT spiked.

TABLE 9.1

SINGLE-BLIND, SPIKED RESULTS FOR TRIBUTYLTIN IN FISH BY GC-FPD. SUMMARY OF TRIBUTYLTIN PRESENT IN FISH AND SHELLFISH BY GC-FPD AND GC-DCP DETECTION, SEPARATELY ON THE SAME SAMPLE

Fish species	TBT added (ppb)	TBT recovered ± SD (ppb)	% Recovery
Flounder	46.84	45.62 ± 1.08 ($n = 3$)	97.4
Flounder	93.68	86.79 ± 3.42 ($n = 3$)	92.6
Whiting	187.36	184.8 ± 3.60 ($n = 3$)	98.6

[a] Single blind protocol was followed; levels spiked were unknown to the analyst doing the tributyltin determinations. Values were compared through a third party.
Reprinted with permission of the copyright owner, Longman Group, UK and *Journal of Applied Organometallic Chemistry* [53].

Figure 9.9 (conditions indicated) illustrates a GC-DCP chromatogram for a typical fish (smoked salmon from Denmark) extract containing high levels of TBT (ca. 825 ppb). Figure 9.10 (conditions indicated) illustrates another typical GC-DCP chromatogram for a mussel (Maryland, USA) extract, at very low levels (33 ppb). Thus, using the sample work-up procedure developed here specifically for megabore GC columns, without prior analyte derivatization, but with direct injection of cleaned-up extracts, peak shapes and chromatographic performance factors by either FPD or DCP detection have proven extremely useful and practical for real world samples. This was one of the few times

TABLE 9.2

SUMMARY OF TRIBUTYLTIN PRESENT IN FISH AND SHELLFISH BY GC-DCP, SEPARATE INJECTIONS ON THE SAME SAMPLES

Fish species (origin)	GC-DCP found [a] ± SD (ppb)	GC-FPD found [a] ± SD (ppb)
Smoked salmon (Denmark)	801.9 ± 58.1 ($n = 9$)	824.5 ± 111.9 ($n = 9$)
Mussel (USA/Maryland)	32.88 ± 2.3 ($n = 6$)	34.03 ± 0.56 ($n = 3$)
Whiting[b] (Uruguay)	185.3 ± 4.7 ($n = 3$)	184.8 ± 3.6 ($n = 3$)

[a] Determined in terms of tributyltin present, rather than tin.
[b] Single blind, spiked sample, no tributyltin incurred. GC-FPD recovery = 98.6%, GC-DCP recovery = 98.9%.
Reprinted with permission of the copyright owner, Longman Group, UK and *Journal of Applied Organometallic Chemistry* [53].

258

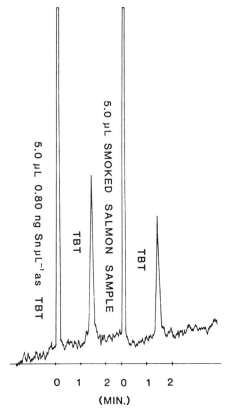

Fig. 9.9. GC-DCP chromatogram for an extract of smoked salmon from Denmark and standard injection of tributyltin, with tributyltin peak as indicated. Specific conditions: megabore 1 μm DB-17 column, 30 m × 0.53 mm, operated at 150°C, injection port 180°C, column flow rate 35 ml min^{-1} argon, DCP at 303.41 nm, gain setting 30, PMT 8, sleeve pressure 50 psi (345 kla), active diagnostic mode, recorder chart speed 0.5 in min^{-1} (1.3 cm min^{-1}, 10 mV FSD). (Reprinted with permission of the copyright owner, Longman Group, UK and *Journal of Applied Organometallic Chemistry* [53].)

that both standards and real world samples had been analyzed by capillary (megabore) GC without prior analyte (TBT) hydridization or alkylation.

Direct injection of an appropriately worked-up sample extract has been very feasible and practical, providing highly accurate and precise quantitation at all incurred levels, down to about 10 ppb. Though our detection limits by both FPD and DCP have been indicated as 50 ppb for a standard solution, in actual fish samples, detection limits were below about 5 ppb. This was because of an effective, overall precon-

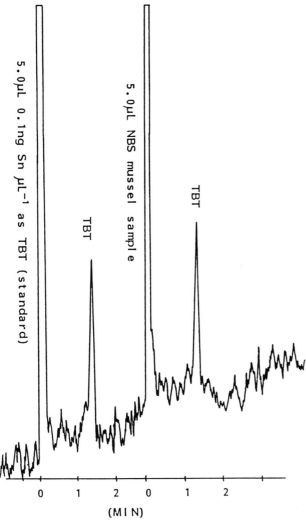

Fig. 9.10. GC-DCP chromatogram for an extract of shellfish (mussel) from Maryland, USA, with tributyltin peak as indicated. Specific conditions: megabore 1 μm DB-17 column, 30 m × 0.53 mm, operated at 150°C, injection port 180°C, column flow rate 35 ml min^{-1} argon, DCP at 303.41 nm, gain setting 40, PMT 9, sleeve pressure 50 psi (345 kla), active diagnostic mode, recorder chart speed 0.5 in min^{-1} (1.3 cm min^{-1}, 10 mV FSD). (Reprinted with permission of the copyright owner, Longman Group, UK and *Journal of Applied Organometallic Chemistry* [53].)

centration of the TBT as a result of extraction. We have been able to accurately and precisely detect TBT at levels in fish or shellfish as low as 13 ppb [53].

References pp. 282–287

9.4. HPLC/DCP FOR TRACE METAL SPECIATION

9.4.1. Background and discussion

Of the work that has already been reported in HPLC-DCP interfacing, certainly that of Uden and colleagues, as above, has been most notable [36, 37]. They have discussed at length some of the advantages possible in using HPLC-DCP, such as its high stability to quite high HPLC flow rates. There is also excellent operational performance with normal phase, organic type solvents for the separations, such as hydrocarbons and halocarbons. It has been possible to obtain linear calibration plots for many metal chelates and derivatives down to the low ng/peak levels. This suggested the possibility that HPLC-DCP detection limits might be comparable with direct-DCP MDLs, but this had not been demonstrated until quite recently [9]. Uden's group has also demonstrated the applicability of paired-ion HPLC procedures for the analysis of cationic surfactants in fuel oils, now using an ion pairing reagent which incorporates specific tagging elements, such as boron in order to achieve high sensitivity. These same workers have also used HPLC-DCP for the separation and characterization of various diethyldithiocarbamate derivatives of Cu(II), Ni(II), and Co(III). Such derivatives were separated on a silica gel analytical column with a mobile phase of acetonitrile/diethyl ether/Skellysolve B, followed by simultaneous UV and DCP detection. Relatively little work has yet been reported that incorporates an initial separation of metal ions, followed by an on-line derivatization to form chelates or complexes, with such derivatives then directly introduced on-line, into the DCP [50, 66–69]. We have undertaken such studies of late, in efforts to improve MDLs for inorganic Se species via paired-ion, RP-HPLC followed by both UV and DCP detection, as below [50].

Some work has been reported on the use of HPLC-DCP for inorganic anions, such as sulfate, nitrate, and acetate, as their cadmium salts. However, the MDLs realized for such metals were in the low ppt (thousand) range via HPLC-DCP, clearly impractical for real world sample applications [2].

9.4.2. Trace analysis and speciation for Cr(VI) and Cr(III) via HPLC–direct current plasma emission spectroscopy (HPLC-DCP)

We have described an HPLC-DCP approach for the analysis and speciation of certain chromium (Cr) ions, especially Cr(III) (chromic) and

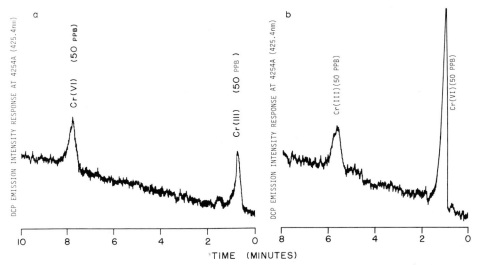

Fig. 9.11. A. HPLC-DCP chromatogram of a mixture of Cr(III) and Cr(VI) standards at the 5 ng/100 μl (50 ppb) injection levels. HPLC used a 15 cm × 4.6 mm i.d. 5 μm C-18 column with mobile phase of PIC A at 2.0 ml min^{-1} flow rate; B. HPLC-DCP chromatogram for a mixture of Cr(III) and Cr(VI) standards injected at 50 ppb levels. HPLC-DCP conditions used a 15 cm × 4.6 mm i.d., 5 μm C-18 reversed phase column with mobile phase of 0.005 M camphor sulfonate counterion at 1.5 ml min^{-1} flow rate. DCP emission monitored at 4254A. (Reprinted with permission of the copyright holder, Preston Publications, Inc. and *Journal of Chromatographic Sciences* [70].)

Cr(VI) (chromate)[70]. Figure 9.11 indicates a typical HPLC-DCP chromatogram for these two Cr containing ions, utilizing an ion-pairing reagent with a commercial reversed phase column for the separations indicated. Element selective chromatograms appear as normal chromatograms, just as one might obtain with more conventional ultraviolet, fluorescence, or electrochemical methods. These HPLC-DCP speciation methods, using two different ion-pairing conditions, have provided detection limits in the low (5–15 ppb) range, which has been ideal for real-world sample applications. These have been, in the absence of hydride generation post-column, electrothermal vaporization, or pre-concentration, some of the lowest MDLs yet reported for any direct HPLC-DCP interfacing. Two separate ion-pairing mobile phase conditions were required in order to accurately and precisely quantitate each Cr species possibly present. This was because these particular ions were of opposite charges (+/−), and one would always be retained and the other unretained, depending on the nature of the ion-pairing agent utilized. However, this has only meant that two separate injections were required in order to fully identify and quantitate both Cr species in any particular sample. Obviously, if only one ion were of interest, then only one form of paired-ion, RP chromatography would be required in HPLC-DCP studies.

Such approaches have been already applied to samples such as: biological samples from drilling locations in the Georges Bank area of New England; soil samples from a chemical dumpsite area known to contain high chromium levels; surface and well-water samples from another chemical dumpsite area outside of Boston, MA; and industrial waste processing samples from a tannery in Maine, known to contain high total chromium levels. These approaches appear to provide suitable MDLs for the speciation of a large variety of chromium containing ions, chelates, complexes, or organochromium derived materials. However, none of the actual samples thus far analyzed by FDA in the mid-80s were found to contain uncomplexed, free Cr ions, such as chromic or chromate. It was also the case that when one intentionally spiked one form of Cr, sample dependent, there was a rapid interconversion to the other ionic or other perhaps complexed forms altogether. High levels of Cr species could be found in certain industrial samples, such as residue from a local tannery after oxidation/ashing. However, actual air samples were never studied via these HPLC-DCP techniques due to a lack of readily available samples. It should perhaps be re-emphasized that the interface always used for these and other HPLC-DCP interfacings, in the absence of post-column hydride generation, as below, used nothing more than a direct connection of the stainless steel or flexible tubing outlet from the end of the HPLC column to the flexible, plastic inlet tubing on the DCP spray chamber [70]. DCP inlet tubing of various diameters was tried in order to maximize peak heights and minimize post-column variances or band broadenings. However, aside from such simple experiments with the tubing connections, there was absolutely no need for any complicated interface between the HPLC and DCP. The DCP is quite stable to both organic and aqueous based mobile phases, as well as having the ability to handle hydride derivatives formed on-line. At times, it also displays excellent overall MDLs, but this is apparently very species dependent.

Perhaps we should discuss the question/problem of MDLs via HPLC-DCP, and why these have only been ideal (5–15 ppb) for Cr containing species, and not general for other metal species under similar/identical HPLC-DCP conditions. That is, our very first efforts involving direct HPLC-DCP for Cr species, as above, provided us with MDLs that have *never* been equaled or surpassed for other Sn or Se containing species, as below [48, 50]. Indeed, there do not appear any other references in the literature to direct HPLC-DCP MDLs that compare with these for Cr species. We had thought, as below, that ion-pairing concentrations, critical micelle formation, or other secondary chemical equilibria in the HPLC mobile phases, might have influenced (improved) inorganic ion transfer to the DCP plume region and hence lowered MDLs. However, this has not been found to be general for any other metal containing species, though serious efforts have been applied for both Sn and Se derivatives. It is possible that only Cr species, for reasons as yet unknown, demonstrate this improved ability to be transported to the DCP plume (enhanced analyte volatility) as a result of secondary chemical equilibria in the mobile phase. However, it is not even clear that such secondary mobile phase effects were ever responsible for the initially observed low MDLs in the above Cr studies. There seems little doubt that these particular MDLs were real, for they have been repeated several times, by the same group of workers. Others have not yet reported comparable results for these particular Cr species under any HPLC-DCP conditions [3, 71].

9.4.3. Continuous hydride generation with direct current plasma emission spectroscopic detection for total arenic determinations (HY-DCP)

The above results for direct HPLC-DCP interfacing and Cr speciation were encouraging, but were not immediately applicable to Sn containing compounds. That is, initial efforts designed to implement HPLC-DCP for Sn species could not result in detection limits below 1–5 ppm, at best [48]. Despite many efforts at varying the nature and concentration of the ion-pairing reagent used in gradient elution RP-HPLC, there was no significant improvement in MDLs beyond the low ppm range. Thus, it became too apparent that some type of post-column, on-line hydride formation step would be necessary in order to lower MDLs to the low ppb range, applicable to real world samples possibly containing organotins. Initially, we undertook some preliminary studies involving direct HY-DCP methods, without any prior chromatographic separation, in order to demonstrate that continuous hydride formation with DCP detection would be applicable for on-line, post-column HPLC interfacing [46]. This was not a flow injection analysis approach, but rather a continuous method, wherein a solution of the sample containing tin species was interfaced with the reagents needed to form the hydride in a continuous manner prior to DCP introduction (Fig. 9.12). The DCP

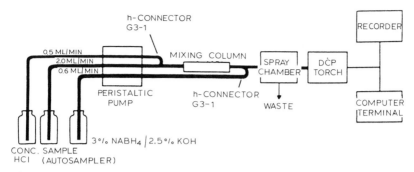

Fig. 9.12. Schematic diagram of the continuous hydride formation–DCP detection instrumental system in actual operation. (Reprinted with permission of the copyright holder, Marcel Dekker, Inc. and *Analytical Letters* [46].)

was configured to receive the aqueous eluent using the normal spray chamber, and the volatile hydrides of Sn were separated (phase separation) in the chamber with high transfer efficiency to the plasma region. In the final method, a simple peristaltic pump hydride formation device was constructed, which utilized conventional, chemical hydride forming reagents (sodium borohydride/hydrochloric acid) prior to mixing with the aqueous sample solution, on-line, in real-time (no time delays). Following system optimization in terms of reagent concentrations, flow rates, mixing time, mixing chamber/plumbing, and DCP operating conditions, analytical figures of merit were derived for detection limits (10 ppb), calibration plots and linearities (100–25,000 ppb, correlation coefficients of 0.999 or better), and so forth. A number of single blind, spiked tunafish samples were analyzed for total arsenic content using the optimized HY-DCP method. A direct comparison was performed of the HY-DCP results for tunafish samples with continuous HY-FAA, and eventually with sequential HY-FAA methods. All three of these methods were directly compared with regard to overall accuracy and precision for total As contents. The methods gave very comparable quantitative results, showing comparable accuracies and precisons. The study showed that this somewhat newer continuous HY-DCP approach for total As determination was an extremely valid, accurate, precise, and reproducible method of trace analysis for actual samples. In addition, the final results suggested that the continuous HY-DCP approach provided percent recoveries consistently closer to 100% (theoretical), in almost all instances. In view of the overall benefits possible by plasma emission spectroscopic detection, either DCP or ICP, a continuous HY-DCP approach, especially one that is now automatable, should be considered highly attractive for future applications. This would be especially true where more than a single element is routinely determined in a very large number of samples, assuming that these can all form hydrides or be introduced into the DCP along with now-formed hydrides from other elements/species. Furthermore, advantages accrue to HY-DCP methods where interferences may be problematic in an FAA approach, and/or where overall automation of such analyses is desirable or required because of the time and costs involved/incurred by alternative analytical methods.

9.4.4. Trace analysis and speciation for methylated organotins by HPLC–hydride generation–direct current plasma emission spectroscopy (HPLC-HY-DCP)

In view of the impractical MDLs via direct HPLC-DCP interfacing for tin species, and the significant improvement in such detection limits for As compounds via HY-DCP methods, it was obvious that an on-line, continuous HPLC-HY-DCP approach should/would prove ideal for perhaps all hydride forming elements [48]. A drawback to the more general applicability of on-line hydride formation is the well-known fact that only a limited number of metals/elements are capable of forming either hydrides or cold vapor Hg derivatives [41, 49]. In order to demonstrate the overall suitability of continuous hydride derivatization-DCP for total tin analysis, we first optimized and applied HY-DCP methods, as above, but now for Sn(II) and three typical organotins, viz., mono-, di-, and trimethyltin. These species were chosen because of a suspicion and interest on the part of the U.S. FDA that they might be present in certain foods destined for human consumption. Following this optimization of HY-DCP for Sn species, a paired-ion HPLC separation was developed for these same organotins, now using an acidic mobile phase with a polystyrene-divinylbenzene copolymer column (PRP-1). The HPLC separation was then interfaced with the HY-DCP derivatization-detection step, and the overall HPLC-HY-DCP approach was optimized and evaluated (Fig. 9.13). Finally, the two procedures (separation-detection) were applied to a large number of spiked and actual samples, including distilled water, seawater, clam juice, clams, and tunafish. The overall results have demonstrated the final suitability, reproducibility, accuracy, precision, reliability, and general ease-of-use of HY-DCP or HPLC-HY-DCP in the analysis and/or speciation for inorganic tin and methylated

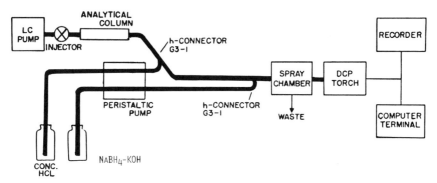

Fig. 9.13. Schematic diagram of the overall HPLC-HY-DCP apparatus for organotin speciation. (Reprinted with permission of the copyright owner, Society for Applied Spectroscopy and *Applied Spectroscopy* [48].)

organotins. It was firmly suggested that the overall method would be applicable to other organotins, such as tributyltin, dibutyltin, and so forth.

In the case of continuous HY-DCP detection for total Sn determinations, it was found that an approximate ten-fold lowering of the MDLs (10–25 ppb) for all three methyltins could be realized in comparison with direct-DCP (100 ppb) methods. A linearity study for stannous fluoride showed a range of 25–10,000 ppb, or about three orders of magnitude, with a correlation coefficient of 0.999. Calibration plots for the three methyltins showed similar ranges of linearity. In the case of HPLC-HY-DCP, final MDLs for all three organotin species could be reduced by at least one order of magnitude (ten-fold) when compared with the HPLC-DCP method. This was what one would have expected based on the initial results obtained when comparing direct-DCP with HY-DCP techniques for total Sn determinations. MDLs via HPLC-HY-DCP for the organotins were in the range of 25–150 ppb, depending on the analyte's retention time, a function of the specific chromatographic system.

Although we had optimized the hydride derivatization step, by varying the concentration of reagents, flow rates of reagents and HPLC eluent, pH effects, and related operating parameters, it was possible that a less than 100% conversion of each tin species to its hydride had occurred. That is, we had not actually measured the percent conversion of each tin species after HPLC separation, nor did we determine how much of this new hydride species was then transferred to the DCP plume for final measurements in HPLC-HY-DCP. Thus, though we have suspected a 100% conversion/transfer efficiency, this was not physically demonstrated. Thus, it is possible, though not very likely, that further optimization of the hydride conversion step and transfer of the hydrides to the DCP plume region might further lower the MDLs. Calibration plots and linearities were performed for all three organotins by HPLC-HY-DCP, and again showing linearities from MDLs for about three orders of magnitude concentration, with correlation coefficients of 0.999. It remains possible that the final responses would be linear beyond the maximum concentration levels studied thus far, but most applications for Sn speciation would not contain such high levels.

Figure 9.14 illustrates a series of three HPLC-HY-DCP chromatograms using the specific HPLC conditions indicated, with 200 μl injections of the organotin standards, at the shown concentration levels. The far left chromatogram (C) is for a mixture of three standards at the levels shown, while the far right chromatogram (A) is the same three organotins spiked to a water sample at levels indicated. Other than for the absolute levels injected, these two chromatograms are identical with regard to species injected and HPLC-HY-DCP conditions employed. Trimethyltin is the last eluting species, which has accounted for its broadness and substantially higher MDLs in comparison with the earlier eluting species. It has been possible to reduce this retention time,

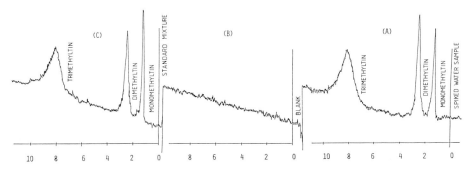

Fig. 9.14. HPLC-HY-DCP chromatograms for the analysis of mono-, di-, and trimethyltin spiked water samples. HPLC column: 25 cm × 4.1 mm i.d., 10 μm, PRP-1, mobile phase 0.003 M 1-hexanesulfonic acid (PIC B-6) + 0.003 M KF +0.02 N H_2SO_4 + 2.5% HOAc, all at 2.5 ml min^{-1} flow rate, 200 μl loop injections, chart speed 0.5 in min^{-1}, full scale = 5 mV. A. Spiked water sample at levels: mono- = 266 ppb; di- = 373 ppb; tri- = 932 ppb. B. Blank. C. Standard mixture at levels: mono- = 300 ppb; di- = 300 ppb; tri- = 850 ppb. (Reprinted with permission of the copyright owner, Society for Applied Spectroscopy and *Applied Spectroscopy* [48].)

and thereby significantly improve its MDL. Gradient elution was not possible in this study, but in the future this might retain overall resolution of all three species while yet lowering the MDL for the last eluter. The repetitious upward slope to the baseline in Fig. 9.12 is, we believe, due to the electronics of the DCP Spectraspan IIIb instrument when operated in the active diagnostic mode. However, this drift did not adversely affect MDLs, linearities, resolutions, or final applications.

The final HPLC separation conditions developed here were a function of several important factors, all of which could be modified at will to alter selectivities and final resolutions, as well as capacity factors. Of crucial importance were items such as pH, acid concentration, ionic strength, flow rate, ion-pairing reagent, concentration of ion-pairing reagent, stationary phase loading, nature of the stationary phase, and other parameters. Peak shape, peak symmetry, column efficiency, alpha values, and final resolutions were all affected or altered by varying these separation conditions. However, the HY-DCP detection conditions, once fully optimized, were always compatible with a wide variety of mobile phase conditions. There was never any indication of an intolerance on the part of the HY-DCP system towards the HPLC separation requirements/conditions. It has always been true that once the DCP or HY-DCP detection steps were fully optimized, univariately, these would eventually be fully compatible with whatever separation conditions were required for full baseline resolution of all metal containing species, as for the organotins here.

The final, fully optimized HPLC-HY-DCP approach for organotins was then validated using single blind, spiked, distilled water samples, containing various types and levels of the methylated tin species. The final, overall results (Table 9.3) indicated very good mass recoveries, ranging from 92 to 106%. These speciation results were further confirmed by determination of the total Sn content in the spiked water samples, via direct-DCP methods. These direct-DCP results for total Sn values agreed quite well with the total Sn values obtained by summation of the individual HPLC-HY-DCP results for each organotin present in the original samples. Even at the 100 ppb levels for the early eluting organotins, percent recoveries were always excellent. Overall, the quantitative results with three separate spiked water samples suggested that the final HPLC-HY-DCP methods of tin speciation were entirely valid, accurate, and precise, at least for artifical samples spiked with known, authentic standards. Additional studies were performed on clam juice, seawater, and tunafish, none of which showed any incurred levels of methylated tins, although many samples were shown to contain various

TABLE 9.3

ANALYSIS OF SPIKED, DEIONIZED, DISTILLED WATER SAMPLES FOR MONO-, DI-, AND TRIMETHYLTIN BY HPLC-HY-DCP [a]

Tin species spiked [b]	Level found (ppb) ± SD (ppb)	Level spiked [c] (ppb)	% Recovery
Monomethyltin	92.8 ± 6.4 (n = 3)	98	94.7
Dimethyltin	105.3 ± 3.7 (n = 3)	107	98.4
Trimethyltin	424.3 ± 2.0 (n = 3)	412	103.0
Total tin by HPLC-HY-DCP [d]	622.4	617	100.9
Total tin by direct HY-DCP	616 ± 10.0 (n = 9)	617	99.8
Monomethyltin	244.2 ± 17.3 (n = 3)	266	91.8
Dimethyltin	365.0 ± 10.5 (n = 3)	373	97.9
Trimethyltin	964.0 ± 40.9 (n = 3)	932	103.4
Total tin by HPLC-HY-DCP [d]	1573.2	1571	100.1
Total tin by direct HY-DCP	1510 ± 21 (n = 9)	1571	96.1
Monomethyltin	627.9 ± 8.7 (n = 3)	634	99.0
Dimethyltin	704.0 ± 16.3 (n = 3)	705	99.9
Trimethyltin	1457.5 ± 23.8 (n = 3)	1376	105.9
Total tin by HPLC-HY-DCP [d]	2789.4	2715	102.7
Total tin by direct HY-DCP	2750 ± 49 (n = 9)	2715	101.3

[a] Sample of deionized, distilled water was spiked, in a single blind study, with all three organotins. The same standards were then used as external standards for HPLC-HY-DCP and direct-DCP quantitative analyses on the same water sample. The analyst did not know any of the Sn levels present in the initial samples.
[b] Standards of Sn species used for spiking were obtained from commercial sources. There was no cross-contamination of standards. All spiked levels were on a Sn basis.
[c] We prepared spiked solutions gravimetrically by weighing out known amounts of each Sn species, and then dissolving in a known, precise volume of deionized, distilled water. Spiked samples were prepared and analyzed within the same working day, together with external quantitative standards prepared and used on that same day.
[d] HPLC conditions used a 25 cm × 4.1 mm i.d., 10 μm, PRP-1 column with 200 μl loop injections, and a mobile phase of 0.003 M 1-hexanesulfonic acid + 0.003 M KF + 0.02 N H_2SO_4 +1% HOAc, all at 3.0 ml min^{-1} flow rate.
Reprinted with permission of the copyright owner, Society for Applied Spectroscopy and Applied Spectroscopy [48].

incurred levels of total Sn. This may have been due to inorganic Sn or alternative organotin species present in such samples, naturally occurring. Single blind, spiking studies were again reported for these real world samples, with percent recoveries ranging from 85 to 102%. Once again, there was very good agreement between spiked levels and those found via HPLC-HY-DCP methods.

References pp. 282–287

In summary, one important advantage of these newer approaches over those previously described, was the ability to derive practical and usable detection limits that were directly and immediately applicable to actual samples. Such approaches have been entirely on-line, real-time, and continuous, and with automated sample injection, could be left unattended for long periods of time. There are serious advantages to using a continuous, post-column, on-line hydride formation step for tin species, as there have been for As species, especially when compared with the older sequential or batch approach to hydride formation. In view of the fact that all of the real-world samples analyzed here for incurred organotins were below our current MDLs, it is important in the future to extend such methods to other Sn containing species and samples, wherein inorganic or organotins have already been detected or are suspect. It has been clear from all the above, that HPLC-HY-DCP speciation methods in the future will or should be based on the use of continuous hydride generation approaches, rather than the batch, discontinuous methods used for so long.

9.4.5. Trace selenium speciation via HPLC with UV and direct current plasma emission spectroscopic detection (HPLC-UV/DCP)

Most recently, we have pursued a program of research and development involving HPLC-DCP applications for Se analysis and speciation [50]. Se is an essential trace element in the human diet, and it falls within the jurisdiction of the FDA to routinely monitor levels of selenate and/or other selenium species or derivatives that may be present in commercial formulations or the natural diet. Unfortunately, there is insufficient data at the moment to conclude what levels and how often selenium (Se) species are to be found in our food supply. It was, in part, the purpose of this particular study to develop improved HPLC-element selective detection (ESD) approaches for Se species in man/animal dietary supplements.

We report here on the direct interfacing of paired-ion, reversed phase HPLC methods for the initial separation of inorganic Se species, selenate and selenite, followed by an on-line, real-time, continuous interfacing with DCP. There was, as before, a direct interfacing via a short, flexible Tefzel connector, which permitted continuous HPLC effluent introduction into the DCP spray chamber at conventional flow rates for inorganic anion separations to be effective. The overall approach, HPLC-DCP, was very similar to what has been reported for Cr and Sn speciation. Optimization of the interface was followed by the derivation of analytical figures of merit for selenate and selenite, as well as by the demonstration of reproducibility of retention times, peak height, peak area, and so forth. Both on-line UV and DCP detection have been utilized, together or alone, for typical separations of designer mixtures of the two inorganic Se species. Single blind spiked sample analyses of selenate/selenite in pure water have been conducted, in order to demonstrate quantitative capabilities prior to real world sample analyses of animal feed premixes. These real samples were then prepared by the FDA laboratory in Denver, spiking selenate and/or selenite at known levels. A comparison of total quantitative levels determined by HPLC-DCP and direct-DCP methods was made in order to further validate the HPLC-DCP approach. Because of the levels of Se containing species normally found in diet supplements for animals, above low ppm values, it has not been necessary to perform any post-column hydride formation prior to DCP detection. This study represented an example of where direct HPLC-DCP interfacing can, at times, provide adequate detection limits (>1 ppm) for analysis and speciation of real world samples. Thus, it is not always necessary to perform post-column, on-line derivatizations prior to final DCP detection for certain samples.

The initial HPLC optimization experiments were performed using UV detection at 205 nm, in order to determine the success of the HPLC separations for selenite(IV) and selenate(VI). Whereas selenate has a low absorptivity at this wavelength, it was adequate to detect selenate at relatively high ppm concentrations. Selenite has a fairly high absorptivity, so that its detection presented no problem. The separation of selenite and selenate could be achieved using quaternary ammonium type ion-pairing reagents under aqueous/organic RP conditions with a C-18 stationary phase. Mobile phases were buffered in a pH range of 6–7 for optimum separations, and methanol or acetonitrile could be used in concentrations up to 10% as the organic modifier. However, because in these studies organics caused higher than usual DCP backgrounds, all HPLC-DCP work used 100% aqueous mobile phases.

Linearity of calibration plots and minimum detection limits (MDLs) were determined by injecting progressively lower concentrations of each species until the detection limit was reached. This was defined as a signal-to-noise ratio of about 2:1, using the absolute background/baseline noise level as reference. In HPLC-UV, both selenite and selenate showed linear responses at concentrations up to about 50 ppm (50 μl injections). Detection limits were on the order of about 0.1 ppm (100 ppb) for selenate and 10 ppb for selenite. Calibration plots were linear over these concentrations ranges, with coefficients of linearity of at least 0.999 or better.

Though MDLs via HPLC-UV were more than adequate for the determination of both Se species in animal feed premixes, it was suspected that interferences might create insurmountable problems by UV detection alone. Especially at 205nm, many interferences from sample components would be expected, possibly to the point of obliterating the analyte peaks. Even with Se standards, considerable baseline disturbance was observed, due to effects of the components of the mobile phase that also respond at 205nm. It was therefore necessary and desirable to investigate DCP detection, concurrent with UV in order to once again illustrate the serious advantages of element selective detection in HPLC. At the primary and strongest emission wavelength for Se (196.026nm), there are no significant interfering elements, and any organic matter would be destroyed by the high temperatures of the plasma without causing wavelength interferences.

In order to optimize the DCP performance in the HPLC interfacing, batch studies were first performed, again using Se standards. Detection limits via direct-DCP were found to be approximately 0.1 ppm (100 ppb), which agreed with the manufacturer's specifications for Se. This work was done in neat aqueous solutions. Since the desired HPLC separation could be realized with no organic modifier present, it was decided to use straight aqueous mobile phases for the FIA- and eventually HPLC-DCP phases.

In the flow injection analysis (FIA)-DCP mode, several parameters were varied for the HPLC separation conditions, to determine their effect on the DCP Se response. Linearity and MDLs were determined for both Se(IV) and Se(VI) in the FIA-DCP mode. Both species yielded linear responses up to 50 ppm (50 μl injections), and the MDLs were about 0.5 ppm (500 ppb) for each. To determine the optimum flow rate entering the DCP spray chamber, a solution of 10 ppm Se(IV) was injected at flow rates varying from 0.2 to 2.0 ml min^{-1}. The mobile phase here was 5 mMQ5 (pentyltriethylammonium phosphate), with 0.01 M each of potassium dihydrogen phosphate and dipotassium hydrogen phosphate. Injection volumes were all 50 μl. The maximum DCP response was realized at flow rates between 0.4 and 0.6 ml min^{-1}, ideal for the HPLC separations needed. At higher flow rates, DCP responses dropped off, with the response at 2.0 ml min^{-1} being about half that possible at 0.6 ml min^{-1}. Below 0.4 ml min^{-1}, the peak became very distorted. Thus, all subsequent experiments were performed at 0.5 ml min^{-1}.

Studies were then undertaken to evaluate the DCP response for Se species made up in ion-pairing medium, in order to determine if micellar formation of inorganic species might lead to DCP enhancements, as discussed above and elsewhere [72]. A series of mobile phases was prepared, each containing 0.01 M mixed phosphate buffer and 5 mM ion-pairing reagent. The ion-pairing reagents were the Q series, i.e., triethylammonium phosphates with 5–8 and 12-carbon chains as the fourth alkyl group completing the tetraalkylammonium cation. Again, 50 μl of a 10 ppm Se(IV) solution was injected with each mobile phase via FIA-DCP. There was essentially no difference in the DCP responses for Q5–Q8. When Q12 was explored, the DCP response actually decreased by about 10%. Since the desired separation could be achieved by using the shorter chain reagents, and since HPLC retentions could be controlled by adjusting the concentration of the ion-pairing reagent and buffer, one still retained the ability to adjust HPLC parameters in order to realize ideal separations on the C-18 column. It is perhaps significant to realize that ion-pairing reagents, as a function of their chain length and perhaps concentration, can adversely affect the DCP responses. MDLs remained constant or decreased throughout these studies, suggesting that micellar formation for Se species does not lead to improved analyte response via heightened mass transfer to the plume region of the DCP. This was contrary to expectations and earlier results for Cr species, again via HPLC-DCP [70].

The above optimized HPLC separation conditions were next interfaced with optimized DCP operating conditions, realized in the FIA-DCP studies. Figure 9.15 illustrates a typical HPLC-DCP chromatogram for the separation-detection of both standard Se species at the levels and conditions indicated. It was apparent that the desired separation and detection of only Se species could be accomplished.

Table 9.4 summarizes both the HPLC-UV and HPLC-DCP results for the qualitative and quantitative determinations of Se(IV) and Se(VI) spiked into water as a single blind validation of the overall methodology. All of these analyses were performed reproducibly ($n = 4$) on two separate days, and the data represents averages $\pm$ standard deviations for all of the data obtained on each sample. In general, all of the results were in agreement with the known, spiking levels, though at times, percent differences could approach 10%. The vast majority of the results had percent differences that were <5% between determined and spiked levels. Standard deviations ($\pm$SD) were less than $\pm$1.0, which suggested that extremely good precision and reproducibility in all quantitative measurements were realized. Even though the minimum

References pp. 282–287

HPLC-DCP

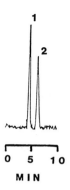

0   5   10

MIN

Fig. 9.15. HPLC-DCP chromatogram for the separation-detection of selenite and selenate species under paired-ion, RP conditions: C-18 column (10 μm, 15 cm × 3.9 mm
i.d.); mobile phase of 2.5 mM TBAHS, 0.01 M each dipotassium hydrogen phosphate
and potassium dihydrogen phosphate, pH = 6.55; flow rate 0.5 ml min^{-1}; DCP detection at 196.026 nm; injections of 50 μl of 20 ppm each Se species as standards.
Peak identities: *1* = selenite; *2* = selenate. (Reprinted with permission of the copyright
owner, The American Chemical Society and the *ACS Symposium Series* [50].)

TABLE 9.4

THE HPLC-UV AND HPLC-DCP RESULTS FOR THE QUALITATIVE AND QUAN-
TITATIVE DETERMINATIONS OF Se(IV) AND Se(VI) SPIKED INTO WATER AS A
SINGLE BLIND VALIDATION OF THE OVERALL METHODOLOGY

Sample	Se(IV)[a]	Actual	% Difference	Se(VI)[a]	Actual	% Difference
HPLC-UV results						
C	12.8 ± 0.1	13.0	−1.5	−	−	−
D	21.4 ± 0.8	22.0	−2.7	−	−	−
E	7.7 ± 0.2	5	+2.7	32.5 ± 0.2	31.5	+3.2
F	−	−	−	15.0 ± 0.5	15.0	0.0
G	24.5 ± 0.1	25.0	−2.0	10.7 ± 0.2	11.0	−2.7
H	−	−	−	9.0 ± 0.6	10.0	−10.0
HPLC-DCP results						
C	12.4 ± 0.4	13.0	−4.6	−	−	−
D	21.8 ± 0.7	22.0	−0.9	−	−	−
E	7.7 ± 0.4	5	+2.7	30.6 ± 0.8	31.5	−2.9
F	−	−	−	16.3 ± 0.7	15.0	+8.6
G	24.4 ± 0.9	25.0	−2.4	11.6 ± 0.3	11.0	+5.5
H	−	−	−	10.4 ± 0.3	10.0	+4.0

[a] Reported as average ppm Se ± standard deviation for $n = 4$, determinations performed
on two separate, consecutive days.
Reprinted with permission of the copyright owner, The American Chemical Society and
the ACS Symposium Series [50].

TABLE 9.5

HPLC-DCP DETERMINATIONS OF SELENATE AND SELENITE IN SPIKED ANIMAL FEED PREMIX, AND DIRECT-DCP DETERMINATIONS OF TOTAL Se CONTENT IN SPIKED ANIMAL FEED PREMIX

Sample[a]	Se(IV)[b]	Actual	% Difference	Se(VI)[b]	Actual	% Difference
1	200.8 ± 9.0	200	+0.4	–	–	–
2	–	–	–	207.2 ± 5.0	200	+3.6
3	105.5 ± 6.5	100	+6.5	113.7 ± 3.2	100	+13.7

Sample[a]	Total Se by direct-DCP[b]	Total Se by HPLC-DCP[b]	Spiked
1	210.8 ± 7.8	200.8 ± 9.0	200
2	212.6 ± 11.2	207.2 ± 3.6	200
3	215.7 ± 3.7	219.3 ± 8.6	200

[a] Quantitations done by matrix matched calibration plots for both Se species, using blank sample containing no spiked/incurred Se content.
[b] Average ± standard deviation, $n = 3$, given as ppm concentrations in feed premix samples.
Reprinted with permission of the copyright owner, The American Chemical Society and the ACS Symposium Series [50].

detection limits by HPLC-DCP were only about 1 ppm, accurate and precise quantitations have been possible in all of these spiked water samples, which ranged from 7.7 to 30.6 ppm Se. Thus, there was no problem apparent for immediate utilization of the newer HPLC-DCP methodology for real world samples. Because of current FDA interest, we selected animal feed premix type samples to further validate the HPLC-DCP method.

These samples were prepared at another FDA laboratory in Denver, CO, wherein known levels of both Se(IV) and Se(VI) species were intentionally added to authentic animal feed premixes. The Denver FDA laboratory did not perform any Se speciation studies. The WEAC laboratory worked-up these samples, and quantitated using a matrix matching technique with a blank feed premix sample spiked at known levels of Se species. Table 9.5 summarizes the data obtained for levels spiked and individual Se speciation. We have indicated average ± standard deviation values for all determinations ($n = 3$). In general, the levels of Se species determined via HPLC-DCP were in good agreement with the actual levels spiked. Table 9.5 also summarizes data obtained for total Se content via direct-DCP methods, together with the summation of HPLC-DCP Se species, along with the total Se species spiked. Again, all of these numbers were in good agreement with each other, especially the results of the HPLC-DCP and direct-DCP total Se values. There was a 5–10% disparity apparent between these values and the actual levels spiked.

It was not possible to perform any qualitative or quantitative determinations via HPLC-UV, because the presence of other UV absorbing species in the elution regions of the Se species prevented their detection (Fig. 9.16). Without further sample work-up and preparation, it was not possible to utilize any form of UV detection with these particular HPLC conditions for either Se species in these samples. In contrast, the HPLC-

HPLC-UV

0   5   10
MIN

Fig. 9.16. HPLC-UV chromatogram of feed premix sample containing nominally 100 ppm Se each as selenite and selenate species. HPLC-UV chromatogram of separation of standard selenite and selenate species (20 ppm Se) under paired-ion, RP conditions: C-18 column (10 μm, 15 cm × 3.9 mm i.d.), RP-300 C-8 guard column (7 μm, 3 cm × 4.6 mm), pre-column Lichrosorb Si60 (30 μm, 2 cm × 3.9 mm); mobile phase of 2.5 mM TBAHS, 0.01 M each dipotassium hydrogen phosphate and potassium dihydrogen phosphate, pH = 6.55; flow rate 0.5 ml min^{-1}; UV detection at 205 nm; injections of 50 μl of 20 ppm each Se species as standards. Peak identities: 1 = selenite; 2 = selenate; 3 = system. (Reprinted with permission of the copyright owner, The American Chemical Society and the *ACS Symposium Series* [50].)

DCP chromatograms for these very same animal feed premix samples were *identical* to a standard chromatogram of both Se species, as in Fig. 9.15, and clearly demonstrated the presence of one or both Se species, sample dependent. A blank injection in HPLC-DCP, of just the sample matrix without spiked selenium species, showed a perfectly flat baseline with absolutely no plasma responses. There was little or no baseline disturbance via DCP detection, as opposed to UV, and the only species peaks observed were at the expected/correct retention times for Se(IV) and/or Se(VI). This was one of the clearest demonstrations of how and when DCP detection provides adequate sensitivity, detection limits, and superb analyte selectivity as well as *specificity* when interfaced with chromatography.

We believe that the above results and chromatograms clearly demonstrated overall method validation for the direct HPLC-DCP interfacing method, and especially the opportunities provided. Unequivocal Se species identification has been possible, along with accurate, precise, and reproducible quantitative determinations in very complex sample matrices. Although the detection limits via direct HPLC-DCP interfacing are, at times, less than ideal, being only about 1 ppm or thereabouts, for animal feed premixes, these are more than adequate and practical. There has not been a need for further pre-concentration of sample extracts, nor for the use of post-column, on-line, chemical generation of more volatile Se hydrides. Future work will, however, attempt to utilize solid phase, post-column hydride formation methods that will permit real-time, continuous, automated, and non-dilution approaches for improved detection limits of these and other Se containing species, inorganic or organometal.

9.5. FUTURE ADVANCES IN CHROMATOGRAPHY/DCP INTERFACING AND APPLICATIONS

It is eminently clear that the DCP instrument, originally designed for batch mode determinations of total metal, can be an ideal element selective/specific detector for many forms of chromatography. Though we have only discussed interfacing with gas and liquid chromatography, the DCP should be compatible with supercritical fluid chromatography (SFC), field flow fractionation (FFF), countercurrent chromatography (CCC), and other separations forms where flow rates are or can be made compatible [73]. It is very clear that in the areas of GC/HPLC, for many modes and flow rates, the DCP can be an almost ideal element selective

References pp. 282–287

detector. It is fully compatible with capillary or packed column GC, as well as preparative, conventional, microbore, or capillary HPLC forms. It also seems very satisfied with normal phase, reversed phase, paired-ion, ion-exchange, ion chromatography, and numerous other forms of HPLC separation modes. It tolerates low or high flow rates, aqueous, organic, or aqueous/organic mobile phases, and organic/inorganic salts in the mobile phases at various concentrations. It can operate in a totally continuous mode for long periods of time, with a very stable baseline, little noise, good signal-to-noise ratios, and generate element selective chromatograms for many elements of importance to the U.S. FDA and other government regulatory agencies. It can provide total element content via direct-DCP or HY-DCP techniques, and is perfectly reliable and stable to an on-line, continuous formation of hydrides from those suitable elements. It can perform under batch hydride, continuous hydride, or flow injection hydride formation conditions, seemingly oblivious to the presence of hydrogen gas, sodium borohydride reaction products, and/or the element's hydrides. When interfaced with HPLC, it can provide ppm detection limits in the direct HPLC-DCP format, and low ppb MDLs when utilizing post-column hydride formation, as in a HPLC-HY-DCP mode. Even when it can only provide low ppm MDLs, via HPLC-DCP, for those elements incapable of forming hydrides, it can still be a very practical and useful approach for speciation in those cases/samples where levels are above 1–5 ppm. In the case of our Se studies, it was not necessary with those particular samples to utilize post-column hydride formation, though this would be entirely feasible and practical based on related work with Sn speciation (HPLC-HY-DCP).

The use of an on-line hydride formation step in the injection port under GC-DCP conditions would permit reaction detection together with GC-DCP determinations [62]. This has already been described for related Sn speciation studies in GC-FPD. It will prove practical for many other elemental species, wherein these are capable of undergoing hydride formation in reaction GC-DCP. Future work should perhaps be directed to improving MDLs in the HPLC mode, since these would appear adequate for >90% of our current samples via direct GC-DCP approaches. In this regard, there are at least three avenues of approach that come to mind. Post-column hydride formation will only work for a certain limited number of elements, less than 10, but when it does work, it works quite well and automatably. Electrothermal vaporization with HPLC-DCP is problematic, in that it still requires a discontinuous, non-automatable, histogram type approach. Fractions are directed to the electrothermal furnace or tantalum ribbon boat, and this must then be desolvated and then vaporized for analyte introduction into the plume region. It is a less-than-ideal approach, and seems more suited to direct-DCP methods, as already described for HPLC-ICP [39]. It has rarely been described, if at all, for HPLC-DCP, and though it would/could provide very low MDLs for all metal species, it would appear today somewhat cumbersome, impractical, and non-continous or automatable. The formation of volatile metal complexes either during or after the HPLC separation, as via post-column chelation with oxine (8-hydroxyquinoline), remains a possible approach [66, 67]. However, this might require some type of complexation, pre-concentration via liquid–liquid extraction, and then phase separation prior to introduction into the DCP [74]. Though the overall scheme seems feasible, it has presented serious problems which still prevent significant lowerings in the final MDLs. Some type of continuous, on-line, eluent/analyte pre-concentration after HPLC separation and prior to DCP introduction is clearly needed. Pre-concentration, pre-column is always feasible, and has been used many times before in direct-DCP or HPLC-DCP modes, but it requires manual manipulation of samples, excessive sample solvent reduction, pre-concentration via removal of solvent or liquid/solid phase extraction, and so forth. It is generally non-automatable, discontinuous, batch-wise, and less than a simple approach to improved MDLs via HPLC-DCP determinations.

Some work has been described for microbore HPLC interfacing with ESD, but most of this has been for ICP/MIP, with relatively little, if any, described with DCP [75]. In principle, it would appear that microbore HPLC could/should provide lower MDLs than conventional bore HPLC-DCP, but this is perhaps misleading. In conventional HPLC, we can inject relatively large volumes of sample, upwards of 200–300 μl are totally feasible. By a judicious choice of elution conditions, most or all of the sample can be pre-concentrated at the head of the analytical column *before* the separation process starts. This effectively can pre-concentrate the analytes into a very narrow band, of perhaps 20–30 μl total volume before elution starts. Thus, it is feasible and has been shown that MDLs can be improved by sample pre-concentration in HPLC using this overall approach. It should also work in HPLC-DCP, though apparently it has not been widely described/practiced. We have used it often in HPLC-ICP, HPLC-HY-ICP, and related approaches. However, in the case of microbore HPLC, one is very limited to the total volume of sample that can be applied to the head of a column before elution begins. This is generally in the range of less than 10 μl, at most, usually less, and even with the same type of pre-concentration approach, the overall benefits are much less than that possible with conventional bore HPLC columns. It is not readily apparent from the literature that microbore HPLC-ESD can or does provide lowered MDLs over conventional HPLC-ESD, wherein the latter is better able to tolerate high injection volumes and analyte pre-concentration before elution. Though microbore or capillary columns are able to provide somewhat improved mass detection limits, it should be clear by now that this does not easily translate into lowered *concentration* detection limits. And after all, it is really lowered concentration MDLs that are crucial when it comes to analyzing real world samples and *not* mass detection limits. We need to detect low concentrations of a species, rather than low masses, in any given sample. The fact that microbore can provide lowered mass detection and improved mass sensitivity, or that the DCP is a mass flow sensitive detector, are not really relevant to the argument or need for lowered concentration MDLs.

Clearly, after reading the other chapters of this text, the final solution to improved MDLs in any form of HPLC-ESD resides in the use of a mass spectrometer as the ultimate/final detection mechanism. Thus, Houk's and Caruso's use of GC/HPLC-ESD/MS for trace metal speciation, at very trace levels (ppt–ppb) has become a most practical, commercial, fully automatable, and routine method of performing true trace metal speciation [41]. Various forms of ESD/MS have been described, and most of these have already been interfaced with HPLC, but that involving DCP seems much less emphasized today. This may be due to the particular configuration of the DCP plasma/plume region, which is not readily adapted as an ion source for mass spectrometry. Thus, though MIP and ICP have been described with MS interfacing, direct and via GC/HPLC sample introduction, there is little, if any, in the literature describing analogous approaches with DCP. There may not be a simple way to overcome this problem, and it may be the case the DCP/MS will not be a very practical alternative to MIP/ICP-MS interfacing, either direct or via chromatography.

There is still, therefore, room for improvement in MDLs for those elemental species not capable of forming hydrides on-line, post-column (HPLC) or pre-column (GC). If ICP/MS continues to develop as it has for chromatographic interfacing, then it will clearly lend itself to solving all or most of the lingering problems in trace metal analysis and speciation with practical samples. It could, in the long run, become the only really useful and practical analytical technique for performing trace metal speciation at very low levels, sub-ppb. It is not clear that GC/HPLC-

DCP is yet able to compete favorably with the latest developments in instrumentation and techniques for GC/HPLC-ICP/MS. This may be a reality and fact-of-life.

9.6. CONCLUSIONS

The DCP instrument can admirably serve as a batch/FIA element selective detector (ESD) for numerous elemental species, providing total metal content for several elements in sequence. In the chromatographic mode, it is an ideal ESD for virtually all volatile metal species via GC, with or without a pre-column, on-line hydride formation step. It works best in a single element mode, as opposed to the current Hewlett Packard GC-MIP unit, which can perform several elemental wavelength determinations simultaneously. Thus, it may perform best when speciation is needed for a single element in its various forms present. It is compatible with virtually all forms of GC, capillary, megabore, packed column, and others. In HPLC areas, it is fully compatible with many separation modes, types of mobile phases, flow rates, and secondary chemical equilibria involved in the overall separation schemes. It is also compatible with on-line, post-column hydride formation steps, and can then really provide useful, practical, trace levels of detection, with true speciation results. However, it can only do this for a limited number of elements. Thus, its practical detection limits for most elemental species are above 1–5 ppm, sometimes satisfactory for particular sample needs, but at most other times, less than ideal for trace applications. At the present time, it will solve some problems very handily and practically, with full automation from sample introduction to data collection. However, when levels of certain elemental species fall below 1 ppm, then HPLC-DCP approaches do not seem viable or practical. There are many alternative elemental speciation methods now available, as described elsewhere in this text, and these could involve some variation of ESD/MS with chromatographic interfacing. Perhaps these are the best of all conclusions to reach?

ACKNOWLEDGEMENTS

Our own efforts in areas of chromatographic interfacing with plasma emission detection, both ICP and DCP, have been materially aided and technically encouraged by several individuals and/or firms: S. Smith, Jr. and R. Schleicher of Instrumentation Laboratory, Inc. (aka

282

Thermo/Jarrell Ash Corp., Franklin, MA), and L. Gershman, K. Panaro, D. Erickson, M. Finkelson, and A. Falco (Winchester Engineering and Analytical Center of the U.S. Food and Drug Administration (FDA), Winchester, MA). Several former graduate stduents at NU were involved in GC/HPLC-AES work, including: S. Jordan, B. Karcher, D. Bushee, and N. Chen. We are most grateful and indebted to all such individuals and any others who have encouraged and/or promoted these efforts.

We are indeed most grateful to the U.S. Food and Drug Administration for providing us with the time, space, instrumentation, materials, finances, and technical support in order to pursue the areas indicated in this review. Some of the FDA work described here was performed as part of the FDA's Science Advisor Research Associate Programs (SARAP). All of our own DCP work was undertaken within the Boston or WEAC laboratories of the U.S. FDA from 1982 to present.

I.S. Krull has been and is currently a Science Advisor to the Winchester Engineering and Analytical Center (WEAC) of the U.S. FDA. Wm. Childress is a research analytical chemist with the FDA at their WEAC facility.

This is publication number 481 from The Barnett Institute at Northeastern University.

REFERENCES

1 I.S. Krull and S. Jordan. GC and HPLC interfaced to plasma emission spectroscopy. American Laboratory, 21 (October, 1980).

2 I.S. Krull. Trace metal analysis by high performance liquid chromatography. In Environmental Analysis by Liquid Chromatography, Ed. by J.F. Lawrence, The Humana Press, Clifton, N.J., 1984, Chapter 5.

3 I.S. Krull. Trace metal analysis and speciation via HPLC-plasma emission spectroscopic detection. Trends in Analytical Chemistry (TrAC), 3(3), 76 (1984).

4 I.S. Krull. Ion chromatography and liquid chromatography in inorganic speciation studies. In The Dahlem Conference on The Importance of Chemical Speciation in Environmental Processes, Ed. by M. Bernhard, F.E. Brinckman, and P.J. Sadler, West Berlin, FRG, September, 1984, Springer-Verlag, Berlin and Heidelberg, December, 1986, p. 579.

5 I.S. Krull. New and potentially novel detectors for HPLC and FIA. In Advances in Chromatography and Separation Chemistry, Ed. by S. Ahuja, ACS Symposium Series, American Chemical Society, Washington, D.C., 1986, Chapter 9.

6 Trace Metal Analysis and Speciation, Ed. by I.S. Krull, Elsevier Science Publishers, Inc., Amsterdam, 1991.

7 Element Specific Chromatographic Detection by Atomic Emission Spectroscopy, Ed. by P.C. Uden, ACS Symposium Series, American Chemical Society, Washington, DC, 1991, in press.

8   The Dahlem Conference on The Importance of Chemical Speciation in Environmental Processes, Ed. by M. Bernhard, F.E. Brinckman, and P.J. Sadler, West Berlin, FRG, September, 1984, Springer-Verlag, Berlin and Heidelberg, December, 1986, p. 579.

9   C.J. Cappon. HPLC speciation of selected trace elements. LC/GC Magazine, 6(7), 584 (1988).

10   L. Colon and G. Barry, Alternating current plasma emission detection in HPLC. In Element Specific Chromatographic Detection by Atomic Emission Spectroscopy, Ed. by P.C. Uden, ACS Symposium Series, American Chemical Society, Washington, DC, 1990, in press.

11   G.W. Rice, J.J. Richard, A.P. D'Silva, and V.A. Fassel. Gas chromatographic-atmospheric pressure active nitrogen method for organomercury speciation in environmental samples. J. Assoc. Off. Anal. Chem., 65(1), 14 (1982).

12   Microwave Plasma Detector Bulletin, MPD 850, Applied Chromatography Systems, Ltd., Luton, UK.

13   Hewlett Packard Technical Bulletins and Application Notes on the HP 5921A Atomic Emission Detector for GC, 1988-90, Hewlett Packard Corporation, Avondale, PA.

14   J.J. Sullivan and B.D. Quimby. Detection of C,H,N, and O in capillary gas chromatography by atomic emission. J. High Resol. Chrom., 12, 282 (1989).

15   D.B. Hooker and J. DeZwaan. Practical applications of a simultaneous AES-MSD GC detector in analytical problem solving. In Element Specific Chromatographic Detection by Atomic Emission Spectroscopy, Ed. by P.C. Uden, ACS Symposium Series, American Chemical Society, Washington, DC, 1991, in press.

16   K.L. Jewett and F.E. Brinckman. The use of element-specific detectors coupled with high performance liquid chromatographs. In Liquid Chromatography Detectors, Ed. by T.M. Vickrey, Marcel Dekker, New York, 1983, Chapter 6.

17   F.E. Brinckman, K.L. Jewett, W.P. Iverson, K.J. Irgolic, K.C. Ehrhardt, and R.A. Stockton. Graphite furnace atomic absorption spectrophotometers as automated element-specific detectors for high-pressure liquid chromatography. The determination of arsenite, arsenate, methylarsonic acid and dimethylarsinic acid. J. Chromatogr., 191, 31 (1980).

18   R.H. Fish, F.E. Brinckman, and K.L. Jewett. Fingerprinting inorganic arsenic and organoarsenic compounds in in situ oil shale retort and process waters using a liquid chromatograph coupled with an atomic absorption spectrometer as a detector. Environ. Sci. Technol., 16, 174 (1982).

19   R.H. Fish and J.J. Komlenic. Molecular characterization and profile identification of vanadyl compounds in heavy crude petroleums by liquid chromatography/graphite furnace atomic absorption spectrometry. Anal. Chem., 56, 510 (1984).

20   R.H. Fish, J.J. Komlenic, and B.K. Wines. Characterization and comparison of vanadyl and nickel compounds in heavy crude petroleums and asphaltenes by reverse-phase and size-exclusion liquid chromatography/graphite furnace atomic absorption spectrometry. Anal. Chem., 56, 2452 (1984).

21   P.C. Uden. Atomic spectral chromatographic detection. In Element Specific Chromatographic Detection by Atomic Emission Spectroscopy, Ed. by P.C. Uden, ACS Symposium Series, American Chemical Society, Washington, DC, 1990, in press.

22   P.C. Uden, Y. Yoo, T. Wang, and Z. Cheng. Element-selective gas chromatographic detection by atomic plasma emission spectroscopy. Review and developments. J. Chromatogr., 468, 319 (1989).

23 D.J. Mazzo, W.G. Elliott, P.C. Uden, and R.M. Barnes. The design and character-
    ization of an interface for high performance liquid chromatography with direct
    current argon plasma detection. Appl. Spec., 38(4), 585 (1984).

24 P.C. Uden. Element-selective chromatographic detection by atomic emission spec-
    troscopy. Chromatography Forum, 17 (Nov/Dec, 1986).

25 C.M.H. Fairless. Studies on the industrial analytical utility of a direct current
    discharge plasma. Amer. Lab. (March, 1978).

26 D.C. Bankston. Processing data from the plasma echelle emission spectrometer.
    Amer. Lab., 31 (March, 1981).

27 A.T. Zander. DC plasma emission spectrometry speeds elemental analysis. Indus-
    trial RandD, 146 (February, 1982).

28 J. Reednick. A unique approach to atomic spectroscopy. High energy plasma exci-
    tation and high resolution spectrometry. Amer. Lab., 53 (March, 1979).

29 Plasma-Spec. The next generation in plasma spectrometry. Technical Bulletin of
    Leeman Labs, Inc., Lowell, MA.

30 T.R. Gilbert and K.J. Hildebrand. A graphite filament plasma/echelle spectrometer
    system. Amer. Lab., 72 (February, 1982).

31 W.R. Biggs, J.T. Gano, and R.J. Brown. Determination of polyphosphate distri-
    bution by liquid chromatographic separation with direct current plasma-atomic
    emission spectrometric detection. Anal. Chem., 56(14), 2653 (1984).

32 I.T. Urasa and F. Ferede. Use of direct current plasma as an element selective
    detector for simultaneous ion chromatographic determination of arsenic(III) and
    arsenic(V) in the presence of other common anions. Anal. Chem., 59, 1563 (1987).

33 V.D. Lewis, S.H. Nam, and I.T. Urasa. Speciation of trace metals by ion chromatog-
    raphy with element selective detectors. J. Chrom. Sci., 27, 468 (1989).

34 I.T. Urasa and S.H. Nam. Direct determination of chromium(III) and chromium(VI)
    with ion chromatography using direct current plasma emission as element-selective
    detector. J. Chrom. Sci., 27, 30 (1989).

35 I.U. Urasa, V.D. Lewis, J. DeZwaan, and S.E. Northcott. Characterization and
    purity determination of trans-($\pm$)-1,2-diaminocyclohexane platinum(IV) tetrachlo-
    ride using liquid chromatography with a platinum selective detector. Anal. Lett.,
    22(3), 579 (1989).

36 P.C. Uden and I.E. Bigley. High pressure liquid chromatography of metal di-
    ethyldithiocarbamates with UV and DC argon-plasma emission spectroscopic de-
    tection. Anal. Chim. Acta, 94, 29 (1977).

37 P.C. Uden, I.E. Bigley, and F.H. Walters. The separation of geometrical isomers
    and mixed ligand forms of cobalt(III) and chromium(III) beta-diketonates by high
    pressure liquid chromatography. Anal. Chim. Acta, 100, 555 (1978).

38 C.M. Kirkman, C. Zu-ben, P.C. Uden, and W.J. Stratton. Developments in the
    high performance liquid chromatography of metallo-organic compounds. J. Chro-
    matogr., 317, 569 (1984).

39 W. Nisamaneepong, J.A. Caruso, and K.C. Ng. Electrothermal vaporization as an
    interface for HPLC introduction to the inductively coupled plasma. J. Chrom. Sci.,
    23, 465 (1985).

40 J.A. Caruso, A. Al-Rashdan, J. Creed, C. Story, D. Heitkemper, and H. Suyani.
    Chromatographic detection by plasma mass spectrometry. In Element Specific
    Chromatographic Detection by Atomic Emission Spectroscopy, Ed. by P.C. Uden,
    ACS Symposium Series, American Chemical Society, Washington, DC, 1991, in
    press.

41   D.S. Bushee. Speciation of mercury using liquid chromatography with detection by inductively coupled plasma mass spectrometry. Analyst, 113, 167 (1988).

42   R.A. Heppner. Elemental detection with a microwave-induced plasma/gas chromatograph-mass spectrometer system. Anal. Chem., 55(13), 2170 (1983).

43   R.S. Houk, V.A. Fassel, G.D. Flesch, H.J. Svec, A.L. Gray, and C.E. Taylor. Inductively coupled argon plasma as an ion source for mass spectrometric determination of trace elements. Anal. Chem., 52, 2283 (1980).

44   R.S. Houk and S.-J. Jiang. Inductively coupled plasma-mass spectrometry for element-selective detection in liquid chromatography. In Trace Metal Analysis and Speciation, Ed. by I.S. Krull, Elsevier Science Publishers, Amsterdam, pp. 87–104, 1991.

45   D.T. Heitkemper and J.A. Caruso. Chromatographic sample introduction for plasma mass spectrometry. In Trace Metal Analysis and Speciation, Ed. by I.S. Krull, Elsevier Science Publishers, Amsterdam, pp. 43–63, 1991.

46   K. Panaro and I.S. Krull. Continuous hydride generation with direct current plasma emission spectroscopic detection for total arsenic determinations (HY-DCP). Anal. Letters, 17(A2), 157 (1984).

47   D.S. Bushee, I.S. Krull, P.R. Demko, and S.B. Smith, Jr. Trace analysis and speciation for arsenic anions by HPLC-hydride generation-inductively coupled plasma emission spectroscopy. J. Liquid Chrom., 7(5), 8671 (1984).

48   I.S. Krull and K.W. Panaro. Trace analysis and speciation for methylated organotins by HPLC–hydride generation–direct current plasma emission spectroscopy (HPLC-HY-DCP). Appl. Spec., 39, 183 (1985).

49   I.S. Krull, D. Bushee, R.G. Schleicher, and S.B. Smith, Jr. Determination of inorganic and organomercury compounds by HPLC-ICP with cold vapor generation. The Analyst, 111, 345 (1986).

50   W.L. Childress, D. Erickson, and I.S. Krull. Selenium speciation in dietary mineral supplements and foods by gas/liquid chromatography interfaced with direct current plasma emission spectroscopic detection (GC/HPLC-DCP). Element Specific Chromatographic Detection by Atomic Emission Spectroscopy, ACS Symposium Series, Ed. by P.C. Uden, American Chemical Society, Washington, DC, 1991, submitted for publication.

51   Chemiluminescence and Photochemical Reaction Detection in Chromatography, Ed. by J.W. Birks, VCH Publishers, New York, 1989, p. 188.

52   I.S. Krull, K.W. Panaro, and D. Erickson. Determination of methylmercury in fish by gas chromatography–direct current plasma emission spectroscopy (GC-DCP). The Analyst, 112, 1097 (1987).

53   K.W. Panaro and I.S. Krull. An improved analytical approach for organotin determinations via GC-flame photometric and direct current plasma emission detection (GC-FPD/DCP). Appl. Organomet. Chem., 3, 295 (1989).

54   S.A. Estes, P.C. Uden, M.D. Rausch, and R.M. Barnes. Fused silica capillary GC separation and element selective microwave plasma emission detection of volatile organometallics. JHRCandCC, 3, 471 (1980).

55   S.A. Estes, P.C. Uden, and R.M. Barnes. Determination of n-butylated trialkyllead compounds by gas chromatography with microwave plasma emission detection. Anal. Chem., 54, 2402 (1982).

56   P.C. Uden and D.E. Henderson. Determination of metals by gas chromatography of metal complexes. The Analyst, 102, 889 (1977).

57   P.C. Uden, R.M. Barnes, and F.P. DiSanzo. Determination of methylcyclopenta-

286

dienylmanganesetricarbonyl in gasoline by gas chromatography with interfaced direct current argon plasma emission detection. Anal. Chem., 50(7), 852 (1978).

58 S.A. Estes, P.C. Uden, and R.M. Barnes. Microwave-excited atmospheric pressure helium plasma emission detection characteristics in fused silica capillary gas chromatography. Anal. Chem., 53, 1829 (1981).

59 K.W. Panaro, D. Erickson, and I.S. Krull. Determination of methylmercury in fish by gas chromatography–direct current plasma atomic emission spectrometry. Analyst, 112, 1097 (1987).

60 I.S. Krull, M. Swartz, and J.N. Driscoll. Multiple detection in gas chromatography. In Advances in Chromatography, Volume 24, Ed. by J.C. Giddings, E. Grushka, J. Cazes, and P.R. Brown, Marcel Dekker, New York, 1984, Chapter 8.

61 "Official Methods of Analysis of the Association of Official Analytical Chemists", Fourteenth Edition, Association of Official Analytical Chemists (AOAC), Washington, DC, 1984. AOAC, 25.150(b).

62 J.J. Sullivan, J.D. Torkelson, M.W. Wekell, A. Hollingsworth, W.L. Saxton, G.A. Miller, K.W. Panaro, and A.D. Uhler. Determination of tri-n-butyltin and di-n-butyltin in fish as hydride derivatives by reaction gas chromatography. Anal. Chem., 60, 626 (1988).

63 S. Clark, J. Ashby, and P.J. Craig. On-column hydride generation method for the production of volatile hydrides of tin, arsenic, and antimony for the gas chromatographic analysis of dilute solutions. The Analyst (London), 112, 1781 (1987).

64 S. Clark and P.J. Craig. The analysis of inorganic and organometallic antimony, arsenic, and tin compounds using an on-column hydride generation method. Appl. Organometal. Chem., 2, 33 (1988).

65 G.A. Junk and J.J. Richard. Tributyltin chloride in water: solid phase extraction, capillary GC separation and EC detection. In Oceans 86 Proceedings, Volume 4, Organotin Symposium, September, 1986, Washington, DC, p. 1160. Copies available from: The IEEE Service Center, 445 Hoes Lane, Piscataway, NJ 08854, USA.

66 B.D. Karcher and I.S. Krull. Fluorescence detection of metal ions separated on a silica-based HPLC reversed-phase support. J. Chrom. Sci., 25(10), 472 (1987).

67 B.D. Karcher, I.S. Krull, R.G. Schleicher, and S.B. Smith, Jr.. On-line extraction of metal ions using liquid–liquid segmentation and a membrane type phase separator for fluorescence detection. Chromatographia., 24, 705 (1987).

68 B.D. Karcher and I.S. Krull, The use of complexing eluents for the high performance liquid chromatographic determination of metal species. In Trace Metal Analysis and Speciation, Ed. by I.S. Krull, Elsevier Science Publishers, Inc., Amsterdam, pp. 105–141, 1991.

69 D. Bushee, I.S. Krull, R.N. Savage, and S.B. Smith, Jr. Metal cation/anion speciation via paired-ion, reversed phase HPLC with refractive index and/or inductively coupled plasma emission spectroscopic detection methods. J. Liquid Chrom., 5, 563 (1982).

70 I.S. Krull, K. Panaro, and L.L. Gershman. Trace analysis and speciation for Cr(VI) and Cr(III) via HPLC–direct current plasma emission spectroscopy (HPLC-DCP). J. Chrom. Sci., 21, 460 (1983).

71 I.S. Krull, D. Bushee, R.N. Savage, R.G. Schleicher, and S.B. Smith, Jr. Speciation of Cr(III) and Cr(VI) via reversed phase HPLC with inductively coupled plasma emission detection. Anal. Letters, 15(A3), 267 (1982).

72 C.M. Kirkman. The application of micellar high performance liquid chromatogra-

phy with specific element detection to the analysis of metal-containing compounds. Ph.D. Thesis, University of Massachusetts at Amherst, February, 1985.

73   C.F. Poole and S.A. Schuette, Contemporary Practice of Chromatography, Elsevier Science Publishers, Amsterdam, 1984.

74   D.S. Bushee, I.S. Krull, S.B. Smith, Jr., and R.G. Schleicher. Post-column, liquid-liquid extraction of organoleads for improved detection in liquid chromatography–ultraviolet detection (LC-UV). Anal. Chim. Acta, 194, 235 (1987).

75   M. Ibrahim, W. Nisamaneepong, and J. Caruso. Microcolumn high pressure liquid chromatography with a glass-frit nebulizer interface for plasma emission detection. J. Chrom. Sci., 23, 144 (1985).

76   Code of Federal Regulations (U.S.), Title 21, Part 573.920; (a), (b1), (b2), (b4), (b5), (c1), April 1, 1986.

SUBJECT INDEX

296

JOURNAL OF CHROMATOGRAPHY LIBRARY

A Series of Books Devoted to Chromatographic and Electrophoretic Techniques and their Applications

Although complementary to the *Journal of Chromatography*, each volume in the Library Series is an important and independent contribution in the field of chromatography and electrophoresis. The Library contains no material reprinted from the journal itself.

Other volumes in this series

300

302